POP ACADEMY

Health and Climate Change

Promoting Health and Wellbeing

Cover Design
Harun Ahmed

First edition

This book was professionally typeset on Reedsy.
Find out more at reedsy.com

Contents

Foreword

"One Health" is an integrative approach that emphasizes the interconnectedness of human, animal, and environmental health, advocating for collaborative, multidisciplinary efforts to address health challenges, particularly amidst globalization and emerging threats. This concept is, no doubt, a challenge to current collective human and institutional behaviors as it shines a spotlight on policies and decisions in human affairs that may often be made without due consideration or recognition of their negative impacts on health outcomes. It also advocates new ways of incorporating health risk assessment into decisions made in a far wider array of private and public sectors than is the current general practice.

While the term "One Health" is perceived to be a modern one, the concept of considering the interconnectedness between human, animal, and environmental health can be traced back to ancient civilizations, with examples like the practices of ancient Greek physicians like Hippocrates who emphasized the importance of a clean environment for human health, ancient Ayurvedic medicine in India which recognized the relationship between animal and human diseases through the study of animal poisons and bites, and Ibn Sina who advocated that the transmission of diseases can be curtailed through proper sanitation.

This integrative approach increasingly recognizes that the health of humans, animals and ecosystems is closely inter-linked and that any changes in these relationships can increase the risk of new human and animal diseases developing and spreading. The close links between human, animal, and environmental health demand close collaboration, communication, and coordination between the relevant sectors.

One Health is an approach to optimize the health of humans, animals, and ecosystems by integrating these fields, rather than keeping them separate. Some 60 percent of emerging infectious diseases that are reported globally come from animals, both wild and domestic. Human activities and stressed ecosystems have created new opportunities for diseases to emerge and spread. These stressors include animal trade, agriculture, livestock farming, urbanization, extractive industries, climate change, habitat fragmentation, and encroachment into wild areas.

Since 2003, the world has seen over 15 million human deaths and US$ 4 trillion in economic losses due to disease and pandemics, as well as immense losses from food and water safety hazards, which are One Health related health threats. According to the World Bank, the expected benefit of One Health to the global community was estimated in 2022 to be at leastUS$ 37 billion per year. The estimated annual need for expenditure on prevention is less than 10 percent of these benefits.

Collaboration across sectors and disciplines through a One Health approach is a vital solution for addressing the complex health challenges facing our society. To prevent, detect, and respond to emerging health challenges, all relevant sectors must collaborate in an integrated manner to achieve together what no sector can achieve alone.

Yet, there are many gaps towards the implementation of 'One Health' and they include inter alia the mapping of existing initiatives and capacities for One Health research and building of the next generation One Health workforce.

This is where Dr. Saroj Pachauri comes into the picture. Together with other experts in the field, she helps to fill in a gap with the publication '"Health and Climate Change: Promoting One Health and Wellbeing". Prof. Dr Saroj Pachauri, a medical doctor, has served in key positions that touch on teaching, training, research, and capacity building while adopting a multidisciplinary approach within the realm of the health sector globally. She is presently the Director and Climate Health Mentor for the POP (Protect Our Planet) Movement. She has received many accolades for her efforts in promoting health and education and is the author of several books and important publications that deal with key issues like the climate crisis and its impact on health and livelihoods.

This 'must-read' present publication brings together many thought leaders and active researchers who share their perspectives with a view to changing mindsets, promoting capacity building, and reforming curricula related to why the world should embrace the concept of One Health especially within the realm of a changing climate.

H.E. Dr. Ameenah Gurib-Fakim
Scientist and Former President of Mauritius

Preface

The serious public health crisis caused by climate change is resulting in several health problems globally. There is an urgent need to undertake research to understand the problems caused by climate change and to design appropriate strategies to address them.

Authors contributing to this book have undertaken research in developed and developing countries to examine these problems. They have suggested approaches to address issues related to One Health, severe heat, air pollution, transgenesis, water dynamics, and plastic pollution.

One Health, an integrative approach, which recognizes the interconnectedness of human, animal, and environmental health, is extensively analyzed and discussed. Efforts undertaken for training and competency development in several European countries are described. The critical role of young public health professionals implementing curricula that emphasize One Health as a cross-curricula subject is discussed.

India's fight against sweltering heat is highlighted. By 2050, heat wave exposure is expected to increase eightfold threatening livelihoods, food security, and public health. Urgent actions needed to strengthen public health systems, improve urban planning, and ensure access to cooling and water resources are discussed. Strategies are suggested to mitigate heat-related risks, especially for vulnerable populations, and recommendations are made to undertake measures to build resilience in the face of a rapidly warming climate.

Air pollution, a significant public health concern in India, is contributing to respiratory and cardiovascular diseases and premature mortality. The government has introduced several action plans including the National Clean Air Program, state and city-specific policies as well as vehicular and industrial emission guidelines. These air pollution measures and their alignment with health policies are examined.

The process of creating transgenic organisms from gene mutation and vector construction to transformation and selection is examined. Among the most beneficial climate transgenic is Golden Rice, bioengineered to combat vitamin A deficiency. Others include genetically modified bacteria producing insulin for the treatment of diabetes and various resistant crops for enhancing food security. However, Monsanto's transgenic maize featuring "terminator technology", a genetic modification for preventing farmers from saving seeds, has sparked ethical and economic concerns. Other potential negative impacts are ecological disruption, pesticide resistance, and health risks. Climate change adds yet another layer of complexity to the transgenic debate. The importance of striking a balance between innovation and ethical responsibility crucial for ensuring that transgenics contribute positively to society and the environment is underscored.

The multifaceted relationship between coastal water dynamics, pollution patterns, and associated health outcomes in coastal communities is examined. A PRISMA review is undertaken to analyze and explore the intricate relationship between coastal water pollution and human health. Management and governance approaches including water quality monitoring and wastewater treatment technologies as well as adaptation strategies for mitigating health risks are examined.

Marine plastic pollution, particularly microplastics, is a growing global crisis that threatens marine environments and contaminates oceans, freshwater, and the food chain. Microplastic accumulation in human tissues raises concerns about chronic diseases such as cancer and cardiovascular problems. Climate change intensifies microplastic pollution by altering ocean currents and accelerating plastic degradation. A comprehensive overview of microplastic pollution is undertaken to analyze sources and pathways to develop effective solutions. Global action is called for to coordinate an effective response to tackle this pressing environmental challenge.

This research from developed and developing countries would be of interest to the academic community, public health scholars, climate change researchers and policy-makers among others.

Acknowledgments

We are most grateful to Komal Mittal, Research Associate, Center for Human Progress, New Delhi, India and Youth Mentor, POP (Protect Our Planet) Movement, New York, USA for her incredible support as we went through multiple iterations of the chapters.

Komal worked untiringly to correct the changes that we made to improve the quality of each chapter. Her pleasant personality and cheerful demeanor made a great of difference to our work on the book. She also finalized the references of several chapters and did the plagiarism checks for all the chapters.

Komal took considerable initiative with the authors of the book chapters which facilitated the process greatly.

1

Synergistic Learning: Embedding One Health in Public Training Curricula and Beyond

Abstract

One Health is an integrative approach recognizing the interconnectedness of human, animal, and environmental health. It emphasizes the impact of zoonotic diseases, environmental changes, and their collective influence on public health. The post-pandemic era underscores the importance of One Health, highlighting zoonosis, the need for cross-sectoral collaboration, and the general public's enhanced understanding of these interdependencies.

ASPHER (Association of Schools of Public Health in the European Region) has historically championed the integration of One Health into public health training. Currently, it spearheads initiatives in climate health and One Health focusing on competency development and the critical role of young professionals in these fields. ASPHER's involvement in a variety of initiatives like COP (Conference of the Parties) and linkages with international groups further illustrates its commitment to these global challenges. The ASPHER-WHO Core Curriculum Programme for Public Health emphasizes the importance of One Health as a cross-curricular subject area. This initiative ensures a broad and inclusive curriculum addressing key areas of One Health competencies.

This chapter provides an overview of competency frameworks which offer a structured approach to embedding One Health principles into education. Academic centers like the UCD One Health Centre play a vital role in this integration focusing on education, surveillance, preparedness, and health literacy. This holistic approach ensures that curricula and societal structures are better equipped to address future public health challenges.

Keywords: One Health, synergistic, curriculum, ASPHER, center

Authors

Karl F Conyard, ASPHER Core Curriculum Fellow, The Association of Schools of Public Health in the European Region (ASPHER), Brussels, Belgium; School of Public Health Physiotherapy and Sports Science, College of Health and Agriculture Sciences, University College Dublin, Ireland; Royal College of Surgeons in Ireland, University of Medicine and Health Sciences, Dublin, Ireland

Tara Chen, ASPHER Climate Health Fellow, The Association of Schools of Public Health in the European Region (ASPHER), Brussels, Belgium; Department of Geography and Environment Management, University of Waterloo, Canada

Uma Divya Kudupudi, Core Curriculum Programme Assistant, School of Public Health Physiotherapy and Sports Science, College of Health and Agriculture Sciences, University College Dublin, Ireland.

Kirsten Duggan, ASPHER Young Professional, The Association of Schools of Public Health in the European Region (ASPHER), Brussels, Belgium

Parnian Jalili, PhD Candidate, School of Public Health Physiotherapy and Sports Science, College of Health and Agriculture Sciences, University College Dublin, Ireland

Jwenish Kumawat, Research Fellow and PhD Candidate, School of Public Health Physiotherapy and Sports Science, College of Health and Agriculture Sciences, University College Dublin, Ireland

Addiena Luke-Currier, PhD Candidate, Department of Sociology, School of Social Sciences and Philosophy, Trinity College Dublin, Ireland

Marie Nabbe, ASPHER Young Professional, The Association of Schools of Public Health in the European Region (ASPHER), Brussels, Belgium; European Hospital and Healthcare Federation (HOPE), Brussels, Belgium

Emrecan Özeler, Public Health Researcher, School of Public Health Physiotherapy and Sports Science, College of Health and Agriculture Sciences, University College Dublin, Ireland

Gerald Barry, Assistant Professor, School of Veterinary Medicine, College of Health and Agriculture Sciences, University College Dublin, Ireland; UCD One Health Centre, College of Health and Agriculture Sciences, University College Dublin, Ireland.

Laurent Chambaud, ASPHER Climate Health Lead, The Association of Schools of Public Health in the European Region (ASPHER), Brussles, Belgium.

Mary B Codd, ASPHER Core Curriculum Programme Lead, Associate Professor of Epidemiology & Biostatistics, School of Public Health Physiotherapy and Sports Science, College of Health and Agriculture Sciences, University College Dublin, Ireland; The Association of Schools of Public Health in the European Region (ASPHER), Executive Board, Brussles, Belgium

Patrick Wall, Professor of Public Health, School of Public Health Physiotherapy and Sports Science, College of Health and Agriculture Sciences, University College Dublin, Ireland

Introduction

One Health is a unique lens through which to view the world we live in, encompassing human, animal, plant, and environment health. The One Health approach is an important viewpoint and is being used internationally by the World Health Organi-

4

zation as an important methodology and management pathway for health for all. Throughout this chapter, One Health will be explored by a variety of means from its importance in a post-pandemic world to how international bodies such as the Association of Schools of Public Health in the European Region (ASPHER) are enhancing and supporting this international process through research, curriculum development, representation, and young professional development. One Health competencies are evolving to meet the changing public health landscape. This chapter offers a practical review of these developments. Finally, this chapter explores how One Health centers at universities connect academia with real-world action supporting policy-makers and professionals in working in the human, animal, plant, and environmental health areas.

Understanding One Health

One Health is increasingly recognized as a unifying approach across public health, environmental, and animal sciences. It is conceptualized as three overlapping spheres connecting human health, animal health, and environmental health. One Health helps to identify and address health and ecosystem threats across the three spheres by recognizing their interconnectedness (1)(2)(3)(4). One main aspect of One Health is its interdisciplinary approach which aims to work across sectors and levels of society (5). Beyond the traditional fields, attention is also being called to the importance of sectors like engineering, business, and the social sciences. All actors engaged in the One Health approach benefit from being cognizant in their mindset in order to attain transformation through collaborative projects (6)(7)(8)(9)(10). One Health, more broadly, utilizes a holistic approach in recognizing the cultural, social, political, and economic factors that shape One Health issues which drive societal change in the way that we view health (9)(11). Historical examples like the return of cholera to the Americans in 1991, the plague outbreak in India in 1994, and the emergence of Ebola in Zaire in 1995 brought awareness to the human health vulnerabilities that are connected with our ecosystems and animal health (12)(13). The way forward to achieve One Health governance is to incorporate diversity in health systems across human health, animal health, and environmental health and to enhance the preparedness of the current infrastructure for existing and evolving health threats across the three spheres (5)(14-16).

Figure 1
One Health

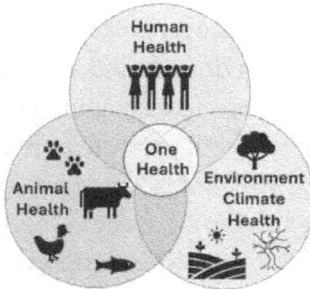

Another way to conceptualize One Health is by examining the interactions depicted in Figure 2. In this figure, the four pillars of One Health—humans, animals, plants, and the environment—are presented. Each pillar plays a crucial and distinct role in safeguarding One Health. The balance of these pillars is essential. If one pillar is weakened or fails to perform its role, it creates an imbalance leading to an increased burden on the remaining pillars. This imbalance can cause heightened vulnerability and potential harm to the entire system, emphasizing the need for each pillar to maintain its strength and function effectively.

Figure 2
The Four Pillars of One Health

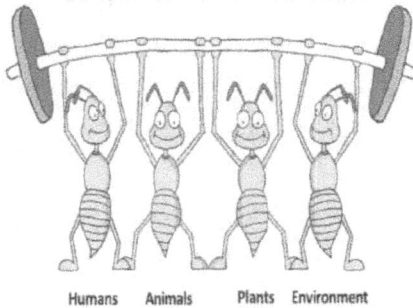

Humans   Animals   Plants   Environment

The concept of positive human health is best understood beyond the formal idea of the "absence of illness." It is primarily sustained by physical, mental, social, economic, and environmental wellbeing (15)(16)(17). A meta-analysis conducted in 2004 to assess the predictors of positive human health identified the relationships between these factors to enable healthcare professionals and researchers to promote best positive health practices (17). Non-communicable diseases effect human health in many ways. Conditions such as cardiovascular disease (CVD), asthma, hypertension, diabetes, and several others have etiological origins in genetic, biological, psychological, cultural, socioeconomic, lifestyle, and environmental factors.

Acute illnesses originate from external factors such as infectious diseases which, in some cases, are caused by unpredictable changes in environmental and animal systems (18). For example, the recent deadly infectious disease COVID-19, zoonotic in origin, quickly spread across the globe through human-to-human transmission (18)(19). Understanding the origins of diseases and their transmission is a key component of One Health. The One Health approach additionally fosters intersectoral solutions to these issues, such as by encouraging urban planning designed to optimize health in the wider community (20)(21)(22)(23).

Figure 3
Human Health

One Health threats to human health are widely recognized and monitored by governments. In the European Union (EU), data collection on serious cross-border threats and infectious diseases is mandated by various binding EU directives and decisions of the European Surveillance System (TESSy), which is maintained by the European Centre for Disease Prevention and Control (ECDC) (24). This infrastructure supports the development of the yearly EU One Health Zoonoses Report which includes data from 38 European countries. It reported that the most common zoonoses in humans in 2022 in this region were Campylobacteriosis and Salmonellosis (24)(25)(26).

Figure 4
Animal Health

Animal health encompasses the mental and physical health of animals as well as the maintenance of production systems to ensure food safety and security (27). The significance of animal health within One Health is particularly recognized within the Quadripartite One Health Joint Plan of Action (2022-2026) (OH JPA) which emphasizes the need to reduce the risks from zoonotic, neglected tropical, and vector-borne diseases (VBDs), address antimicrobial resistance (AMR), and ensure food safety and security (28). Zoonotic diseases are those which can spread between people and animals via direct or indirect contact as well as via food, water, or vectors (29)(30).

VBDs in particular are a critical threat, particularly due to their expansion in geographical range (31). Animal welfare is a crucial factor in animal health as stressed animals are more prone to infectious diseases, increasing their need for antibiotics and the risk of developing antibiotic resistance.

Globally, a large proportion of antimicrobials sold are used in food producing animals and their misuse, particularly in intensive industries, can be linked to resistance and infections in humans and animals which are harder to treat. The threat of AMR is closely linked to food safety due to the potential route of transmission (32)(33)(34). Food safety is a broad concept which encompasses threats along the food supply chain including zoonotic and non-zoonotic pathogens, contamination of food and water by animals and humans, and more (28).

The production of food producing animals can also be linked to the environmental health aspect of One Health as agriculture is a significant contributor to greenhouse gas (GHG) emissions and water pollution. The Intergovernmental Panel on Climate Change (IPCC) estimated in their Assessment Report that between 2010 and 2019, agriculture, forestry and other land use accounted for 12 to 21 percent of global anthropogenic GHG emissions (35). This example demonstrates the complex nature of One Health problems as some solutions for reducing GHG emissions in agriculture have been found to cause negative impacts on human and animal health (36).

The health of the environment and its climate can be described as the ability of the environment to maintain its function and processes and adapt to changes driven by human activity, though it can be conceived more broadly to include the health of ecosystems and related areas (28)(37). There has traditionally been a heavier emphasis within the One Health concept on human and animal health and there has been an identified need for further inclusion of the environmental and climate science sector (38)(39). The OH JPA aims to fully integrate the environmental sector within the One Health approach (28).

There are many environmental concerns within One Health such as environmental degradation, pollution, contaminants, soil and waterborne diseases, and other issues which can pose a threat to human, animal, and plant health (28)(40). Two significant issues include climate change and AMR, particularly as environmental bacteria can be a reservoir for AMR genes (41). Climate change can cause human, animal, and plant health, both directly and indirectly, including heat-related illness, natural disasters, food insecurity, climate migration, the spread of infectious diseases, and the emergence of insect vectors in new regions (42)(43).

Figure 5
Environmental Climate Health

The Lancet Countdown has developed specific indicators for the tracking of climate change and health which continue to be updated (44). New indicators within the European 2024 report recognize the complex interaction of environment and human health by tracking healthcare sector emissions and other related harms and acknowledging its role in impacting key goals such as decarbonization (45).

Post-pandemic importance of One Health

The COVID-19 pandemic has irrevocably transformed our understanding of global health, underscoring the profound interconnectedness between human, animal, and environmental health (46). This interconnectedness, long recognized within the framework of One Health, has gained renewed attention as the world grapples with the aftershocks of a crisis that transcended borders, species, and ecosystems. One Health is not just a concept; it is a multidisciplinary approach that integrates the efforts of multiple sectors to achieve optimal health outcomes, acknowledging that the health of humans, animals, plants, and the environment is inextricably linked.

The pandemic served as a stark reminder of how zoonotic diseases—those transmitted between animals and humans—can have devastating consequences on a global scale. The rapid spread of COVID-19, believed to have originated from a wildlife source, highlighted the urgent need for a more integrated approach to health where human actions, environmental stewardship, and animal health are seen as parts of a single unified system (47). The emergence and re-emergence of zoonotic diseases is not new phenomen but the pandemic has brought them into sharp focus emphasizing the need for vigilant surveillance, early detection, and rapid response strategies that span across sectors and borders.

Figure 6
Virus Particles

As we transition into the post-pandemic world, lessons learned from COVID-19 are shaping the future of public health. The One Health approach, which was once a theoretical framework for addressing health issues, has now become a practical necessity. The pandemic demonstrated that health challenges cannot be effectively addressed in isolation; they require a coordinated effort that includes expertise from public health, veterinary science, environmental science, and beyond. This approach ensures that health strategies are comprehensive, taking into account the complex interdependencies that characterize modern health challenges.

This section delves into the post-pandemic importance of

One Health focusing on three critical areas: zoonoses control, cross-sectoral methodology, and enhanced understanding of the general public of the One Health link. By exploring these themes, we aim to highlight the essential role of One Health in preventing future pandemics by promoting collaborative work across disciplines and fostering a more informed and resilient global community. The COVID-19 pandemic has served as a wake-up call highlighting the need for a more robust and holistic approach to global health—one that acknowledges the interconnectedness of all life on Earth.

Figure 7
COVID Precautions

Zoonosis and One Health

That zoonotic disease pathogens are transmissible from animals to humans has been recognized as a significant threat to global health. These diseases are responsible for some of the most devastating pandemics in human history including the bubonic plague, influenza, and more recently, COVID-19. Approximately 60 percent of all emerging infectious diseases are zoonotic with over 70 percent of these originating from wildlife (30). This statistic underscores the importance of a holistic approach that integrates human, animal, and environmental health to prevent, detect, and respond to zoonotic threats effectively (48).

The role of human activity in zoonosis

Human activities such as deforestation, agricultural expansion, and urbanization, disrupt ecosystems increasing contact between people and wildlife which raises the risk of zoonotic spillover events (49). Global trade in wildlife further exacerbates the transmission of zoonotic diseases. The One Health approach emphasizes the need for addressing these root causes by promoting sustainable practices that protect habitats and minimize human-wildlife interaction, thereby reducing the likelihood of future pandemics and safeguarding global health (50). In addition, with more people owning pets than ever before, the closer relationships between humans and companion animals amplify the risk of zoonotic disease spread.

Figure 8
Zoonotic Connection

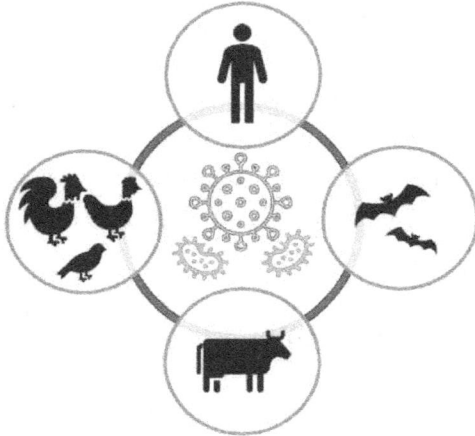

Zoonotic disease surveillance and response

Effective surveillance is crucial for managing zoonotic diseases. Systems that simultaneously monitor the health of
humans, animals, and ecosystems enable early detection
of threats (51). Integrating data from veterinary, environmental, and public health sources allows for a comprehensive understanding of disease dynamics and facilitates rapid
responses crucial for advancing this field. Whole genome
sequencing (WGS) and bioinformatics are revolutionizing
molecular surveillance, offering deeper insights into the
transmission networks of zoonotic agents, and enhancing our
ability to track and respond to outbreaks. Microbes are highly
promiscuous with their genetic material and the exchange
of this material creates opportunities for new pathogens
to emerge and existing pathogens to acquire new virulence

characteristics. WGS enables us to track these mobile genetic elements. A major concern is the potential emergence of an influenza variant with the virulence of highly pathogenic avian influenza (HPAI) and the transmissibility of human influenza.

Figure 9
Avian Influenza Infographic

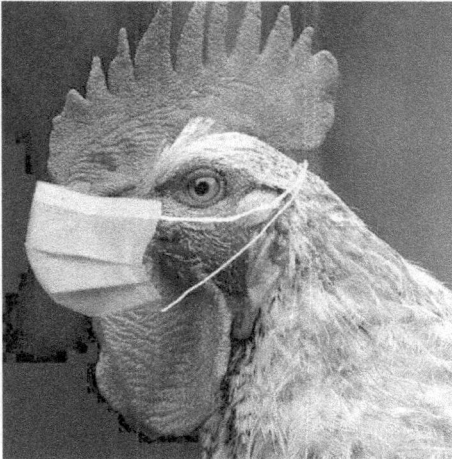

WGS can identify the genetic markers and mutations that signal the development of such a variant allowing for early detection and rapid response to prevent a potential pandemic. Establishing a One Health database that integrates sequenced isolates from humans, animals, food, plants, and the environment with existing open-access databases is a crucial advancement in this field." The COVID-19 pandemic revealed gaps in global surveillance, highlighting the need for stronger international cooperation, improved diagnostics, and enhanced laboratory capacities to better manage zoonotic disease outbreaks (52).

Prevention and preparedness

Preventing zoonotic diseases should be a global health priority, shifting from reactive to proactive measures. The One Health approach emphasizes comprehensive risk assessments that account for human, animal, and environmental health (53). Investing in research for vaccines, diagnostics, and treatments from this perspective is crucial (54). Effective control may involve eradicating or vaccinating animals to prevent zoonoses in humans. Public health policies should be evidence-based and focused on reducing risk at its source while education about health interconnections can promote responsible behaviors and support protective policies.

Figure 10
One Health and NCDS

One Health and Non-communicable Diseases

Non-communicable diseases (NCDs) are increasingly recognized as part of the One Health paradigm, as they often arise from shared environmental risk factors affecting both humans and animals. Animals can serve as sentinels for environmental hazards, offering early warnings of potential risks to human health. For example, the presence of tumours in food animals or wildlife exposed to pollutants or the onset of respiratory issues in pets living in areas with poor air quality can alert authorities to similar risks in human populations. By monitoring animals as indicators of environmental health, we can implement proactive measures to prevent or mitigate the impact of NCDs in humans.

Figure 11
Cross-Sectoral Collaboration

Cross-Sectoral Methodology and Collaborative Work

The One Health approach underscores the interconnectedness of human, animal, and environmental health. Addressing complex health challenges, especially in a post-pandemic world, necessitates cross-sectoral collaboration. The COVID-19 pandemic highlighted the need for coordinated efforts across multiple disciplines rather than isolated sector-specific responses. (55)

The need for collaboration across sectors

Health issues, particularly zoonotic diseases, stem from the interactions between humans, animals, and the environment. Effective management requires collaboration among public health officials, veterinarians, environmental scientists, policy-makers, and other stakeholders. Key reasons for cross-sectoral collaboration include:

Table 1	Requirements for cross-sectoral collaboration
Holistic Understanding	Combining diverse expertise provides a comprehensive view of health threats, enabling better identification and mitigation of disease drivers.
Efficient Resource Use	Pooling resources, including data and technology, avoids duplication and ensures coordinated, effective responses.
Innovation	Interdisciplinary collaboration fosters unique perspectives, leading to innovative and sustainable health interventions.
Informed Policies and Decision-Making	Policies informed by insights across sectors address the interdependencies of human, animal, and environmental health more effectively.

Challenges to effective cross-sectoral collaboration

Table 2	Challenges in cross-sectoral collaboration
Despite the benefits, challenges remain:	
Communication Barriers	Scientific language and methodologies can hinder effective communication. Establishing common terminology and goals is crucial.
Institutional Silos	Independent operation by government ministries agencies and other organizations can create barriers. Overcoming this requires a cultural shift towards interdisciplinary collaboration.
Resource Constraints	Collaborative efforts demand time, funding, and coordination. Adequate support is essential for success. The benefits of prevention through collaborative work are hard to quantify in terms of return on investment.
Political and Institutional Resistance	Competing interests or power dynamics can obstruct collaboration. Strong leadership and a shared commitment to the One Health approach are necessary to address these challenges.

The path forward: BRACE for One Health

To strengthen cross-sectoral collaboration in the post-pandemic world, several strategies should be pursued:

Table 3	The BRACE model for public health
B Building Institutional Support	Establish structures like interdepartmental committees, joint funding mechanisms, and integrated policy frameworks to facilitate ongoing collaboration between sectors.
R Raising Public Awareness	Engage and educate the public about the interconnectedness of health across species and ecosystems, building support for One Health initiatives and encouraging responsible behaviors.
A Advancing Interdisciplinary Research	Increase investment in research projects that involve multiple sectors to better address complex health challenges, promote innovative solutions, and get the sectors working together.
C Capacity Building	Implement training programs that equip professionals with the skills needed for cross-sectoral work, with a focus on communication, teamwork, and systems thinking.
E Enhancing Global Cooperation	Foster international collaboration by sharing data, resources, and best practices to coordinate responses to global health threats more effectively.

Advanced knowledge of the One Health link

The COVID-19 pandemic has created an awareness of the risks of the spread of infectious diseases amongst the public. Before the pandemic, terms like incubation period, viral variants, messenger RNA vaccines, zoonosis and One Health etc. were confined to the realms of experts and academia. But as the crisis unfolded, these concepts became part of everyday conversations—on the news, in homes, and across social media. The pandemic forced us all to see that the health of humans, animals, and the environment are not separate issues but parts of a whole intricately woven together.

Figure 12
Loud Speaker

This newfound public awareness is both a challenge and an opportunity. On one hand, it means that there is now a broader understanding of how our actions—whether it's encroaching on wildlife habitats or misusing antibiotics in agriculture—can have far-reaching impacts. On the other hand, it requires us to build on this awareness to drive lasting change. The public is more informed, yes, but they are also more expectant. They expect leaders to take the lessons of the pandemic seriously to implement policies that protect health in its broadest sense and ensure that the mistakes of the past are not repeated.

Moving forward, this advanced public knowledge of the One Health link must be nurtured and expanded. Education systems, public health campaigns, and community engagement initiatives all have crucial roles to play. We need to continue to empower people with the knowledge that their choices matter—not just for their own health, but for the health of the planet. This isn't just about preventing the next pandemic; it's about building a society that understands and respects the delicate balance that sustains life on Earth.

The pandemic has taught us many hard lessons but perhaps the most important one is that we are all connected—humans, animals, and the environment. The One Health approach is not just a strategy for experts, it's a mindset for everyone. By embracing this perspective, we can create a future where health is truly holistic and where the wellbeing of people, animals, and the planet seen as one and the same.

ASPHER leading the way

The Association of Schools of Public Health in the European Region (ASPHER) has a long history. The organization began operations in 1968 and is continuously striving in its dedicated quest in improving and protecting public health by strengthening education and training public health professionals for both practice and research. Regarding One Health, key work actively carried out by the many teams at ASPHER is presented below.

Figure 13
ASPHER Logo

The Association of Schools of Public Health in the European Region

ASPHER activity for environmental climate health and One Health

The links between humans and nature are at the core of One Health and climate health topics. While the public health workforce may have become much more visible during the COVID-19 pandemic, its competencies have always been crucial in a constantly evolving world but now, with the challenges of climate change, it needs to come into even sharper focus.

With climate change posing a growing health risk, those working in health systems need to be adequately trained to identify, manage and, if possible, prevent the resulting negative health effects and this includes both clinical and public health professionals (57). Both groups "have important roles and responsibilities, some of which are shared, that must work for society to successfully mitigate the root causes of climate change and build a health system that can reduce morbidity and mortality impacts from climate-related hazards"; while at the same time providing a balance between human and animal health (58).

ASPHER climate health competencies

In 2020, ASPHER collaborated with the WHO to create the WHO-ASPHER Competency Framework for the Public Health Workforce in the European Region (59). Within this framework, One Health, together with health security, forms one of the pillars of the category "Content and Context". This section introduces two important competencies, asking that the public health workforce.

Furthermore, following the announcement of the European Green Deal in 2019, ASPHER started surveying member schools in 2020, aiming to establish how they could best support the goals of the Green Deal through their network of public health schools (60). As stated by Orhan et al., "a systematic approach is missing and there is a lack of strategy in most schools" concerning climate actions (61).

In response, ASPHER published the "Climate and Health Competencies for Public Health Professionals in Europe" in October 2021 (61). Building on different lists of competencies from the GCCHE's Core Climate & Health Competencies for Health Professionals, ASPHER adopted their catalogue of competencies adapted to the European Region, divided into four domains (62). Table 4 provides an overview of the domains together with the main focus points (61).

Figure 14

WHO-ASPHER Competency Framework One Health and Competency Frameworks

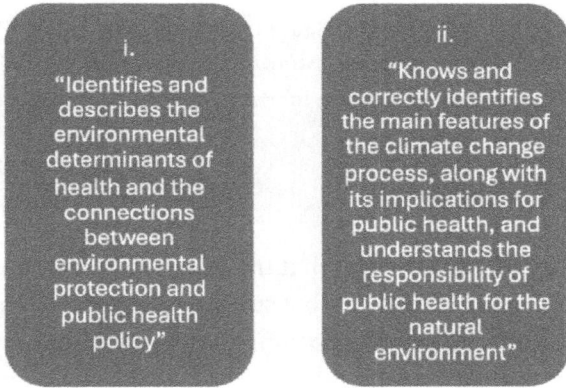

i.

"Identifies and describes the environmental determinants of health and the connections between environmental protection and public health policy"

ii.

"Knows and correctly identifies the main features of the climate change process, along with its implications for public health, and understands the responsibility of public health for the natural environment"

Table 4		The four domains of climate and health competencies for public health professionals in Europe
DOMAINS	Knowledge and Analytical Skills	-Knowledge about and ability to identify drivers of climate change -Knowledge about and the ability to identify climate change-related health impacts as well as effective responses -An understanding of biodiversity/ habitat losses and infectious diseases -Knowledge about preventive actions, climate mitigation, adaptation and their co-benefits -An understanding of the social and environmental determinants of health -An understanding of emergency planning and preparedness -An ability to access and interpret relevant and accurate information/ science about climate change - related health effects from different sources and at different levels -An ability to apply knowledge with the aim of improving decisions in public health services and public health -An ability to develop strategies for a carbon footprint reduction in healthcare delivery -An awareness of the relevant ethical, professional, and legal obligations
	Communication and Advocacy	-An ability to communicate effectively with different stakeholders and to distribute information in audience-appropriate ways -An understanding of the role of the public health workforce in activism and policy engagement regarding climate and health
	Collaboration and Partnerships	-An ability to work collaborate across sectors in different organizational structures/ at different levels on climate and health issues
	Policy	-An understanding of the role of different policy frameworks and governance structures needed to address health risks associated with climate change

Aligned with the importance of developing the framework (61), ASPHER took the lead on the 2021 thematic networks of the EU Health Policy Platform and of the Joint Statement: Moving towards the right to 'health for all' by training the public health and wider health workforce on climate change and health (63) . The Joint Statement states:

Figure 15

2021 EU Health Policy Platform: Joint Statement Quote

"Public health and healthcare professionals require core training and continuous professional development to improve their understanding of the linkages between climate and health and to make it a priority in their work".

It also highlighted the fact that the planetary health and One Health concepts should be—but still are not—central to public health education, creating an unprepared workforce (62). The Joint Statement ended with a call for action on 13 points, urging to bring climate change and health concerns to the forefront of the debates. Following the release of the Joint Statement, the ASPHER Climate and Health Working Group was created in 2022. Among the numerous actions carried out by this group, ASPHER joining the European Climate and Health Observatory as an official member and the creation of the European Climate and Health Responder Course in 2023 in collaboration with the Global Consortium on Climate Change and Health Education (GCCHE), stand out in particular.

Linkage with the Global Consortium of Climate Change and Health Education

True to their vision that "all health professionals throughout the world will be trained to prevent, mitigate, and respond to the health impacts of climate change" the GCCHE released a first iteration of their core concepts in 2018 and have since updated them every 18 months (64). In the creation of the "Climate and Health Competencies for Public Health Professionals in Europe", ASPHER built the competencies on the GCCHE concepts, aiming for strong coherence and harmonization (61). ASPHER closely followed the last update of the GCCHE's "climate health core concepts for health professionals" in 2023 and is in discussion with the GCCHE regarding the next update in the European region (64).

In order to assess the impact of their work, GCCHE collaborated with Schools of Public Health networks such as ASPHER to take stock of climate health courses, content, and programs across the globe. ASPHER's first iteration of the survey was led by Orhan et al. (2024) with future goals to align with the GCCHE as a global data indicator with periodic collection to monitor the evaluation and impact of climate health education efforts.

Figure 16
Columbia MSPH GCCHE Logo

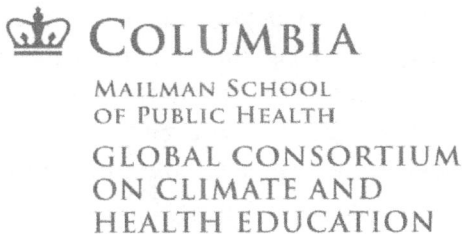

COLUMBIA

MAILMAN SCHOOL
OF PUBLIC HEALTH

GLOBAL CONSORTIUM
ON CLIMATE AND
HEALTH EDUCATION

Starting in 2023, the ASPHER Climate and Health Working Group collaborated closely with the GCCHE to create the European Climate and Health Responder course, an online 10-week course to build climate health literacy among health professionals (65). The course ran from February to April 2024 with the aim of bringing an introduction to the main climate and health issues to health and public health professionals within the European region. The course was well received with participants joining from all over the world, not just Europe. The Climate and Health Working Group is currently working on further courses with the GCCHE.

ASPHER's young professionals' programme

Since 2020, ASPHER has been running the Young Professionals Programme (YPP) (66). Aimed at public health students and early career or "young professionals" (YP), the program encourages them to build and enhance their network and collaboration opportunities while gathering advice and experience. YP are involved in all areas of ASPHER 's work including the Climate and Health Working Group, where they were involved for example, in the creation of the Responder Course and Support Work looking at planetary health and One Health. As the YP tend to join the Climate and Health Working Group due to an existing interest in the topic, they form an asset to ASPHER 's work in spreading information on the need to include these elements in the training of health and public health professionals.

Figure 17
Hands Together

The Conference of Parties (COP)

In 2023, ASPHER received Observer Status for COP 28. The Conference of Parties (COP) Climate Change Conference was held for the 28th time in 2023 and was the first conference with a whole day devoted to the topic of Climate Change and Health (67).

In this context, the "ASPHER statement for COP28: A call for action in seven points" was released, outlining the seven most important steps to help reduce the impact of climate change on humans, ecosystems, and the planet (68). These include, not only the recognition of the link between climate change and health and the need for an integrated global approach to the topic with solidarity between countries, but also stress the need to tackle the effects climate change has on health inequalities as well as the necessity for further advocacy and transdisciplinary research. Furthermore, the statement reinforces the necessity of creating a well-prepared workforce to mitigate the effects of climate change on health. ASPHER continues to hold its presence at policy accelerating events such as COP29 and will continue to actively work on advancing the issues around climate change and health.

Figure 18
COP28 UAE Logo

ASPHER Core Curriculum Programme for Public Health: One Health

The ASPHER Core Curriculum Programme for Public Health (CCP) is a vital curriculum development and capacity building project for schools of public health, trainers, and professionals in the European region. A bottoms-up approach was used to gather consensus from experts in the area of public health academia, public health medicine and practice, alongside young professionals and current graduate students from across the world. The consortium has joined together to assist in the creation of a unified connected curriculum which is able to contend with the changing public health landscape which we call the "New Normal". This New Normal has impacted the health of the public in a variety of ways from climate change, migration, conflict, war, novel infectious disease pandemics, and current crises. One Health plays an integral role as a cross curricular subject area within the CCP. Understanding and embracing the One Health approach has become essential for preparing future public health professionals to tackle complex, interconnected health issues that define our time.

Figure 19

ASPHER Core Curriculum Program

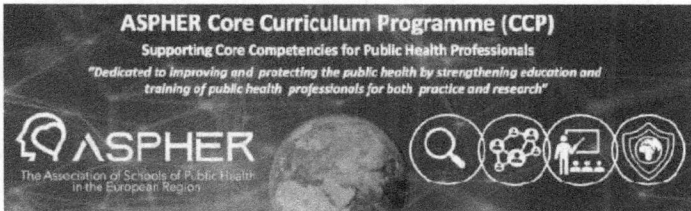

Development of core public health competencies has been core to ASPHER's mission since 2006. ASPHER's Core Competencies Programme was originally led by ASPHER Past President, Professor Anders Foldspang of Aarhus University, Denmark and Dr. Christopher A. Birt of University of Liverpool, England (69). In 2022, it was agreed that a new direction was required in which a shift from competencies to curriculum development to better assist schools of public health in the European region and to assist professional training programs.

The CCP uses a bottoms-up approach involving multiple partners from a variety of different backgrounds from universities, governmental and non-governmental organizations to special occupational interest groups internationally. The CCP methodology is described below.

The impact of the CCP extends through 130 schools and institutes of public health. European European countries. annual ASPHER deans' and directors' retreats since 2022 (Sofia, 2022; Dublin, 2023; Porto, 2024) have been integral in the practical application and development of the CCP by means of organized workshops and educational sessions for delegates. There has been active participation in the curricular content of specialist areas of the CCP by public health professionals, young professionals, and graduate students of public health across Europe and beyond.

A unique feature of the CCP is that the program is supported by over 300 public health professionals across 36 Expert Advisory Groups (EAGs) each representing a separate subject area. EAG membership represents experts in public health/population health academia from many schools of public health across the globe, public health practice experts from across governmental and non-governmental organizations, young professionals in public health, and current students of public health.

Figure 20
ASPHER Core Curriculum Program Methodology

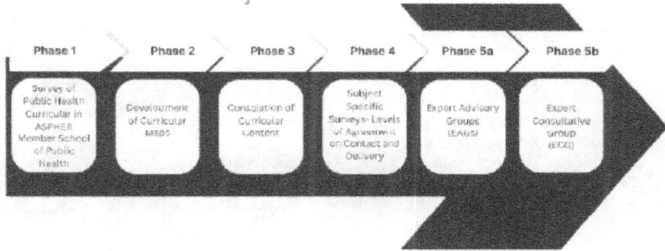

A unique feature of the CCP is that the program is supported by over 300 public health professionals across 36 Expert Advisory Groups (EAGs) each representing a separate subject area. EAG membership represents experts in public health/population health academia from many schools of public health across the globe, public health practice experts from across governmental and non-governmental organizations, young professionals in public health, and current students of public health.

Figure 21
Connection

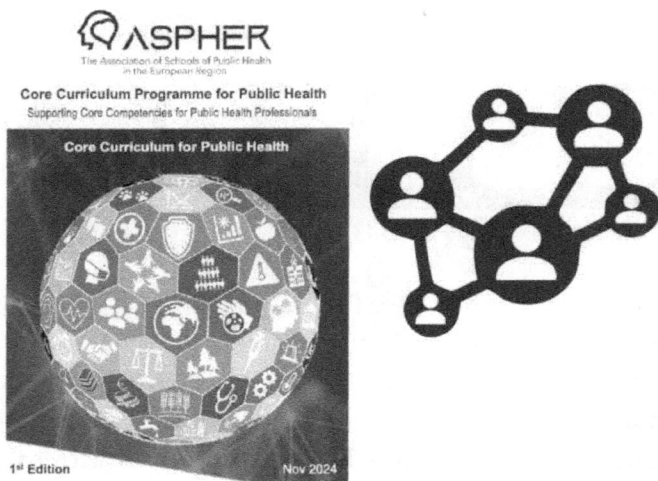

The outcome of this program, conducted from 2022 to 2024, is a comprehensive curriculum for public health in Europe which includes 36 subject areas collated into four domains, each with a function in public health education.

Figure 22
ASPHER CCP Domains

Domain 1 collates subject areas essential in any public health education program (core subject areas), i.e. demography, epidemiology, public health research methods, biostatistics, determinants of health, health protection, disease prevention, health promotion, law for public health, public health ethics, economics for public health, and health systems organization.

Domain 2 (subject-specific areas) are subjects which may be included in general public health programs or specialist streams within programs i.e. communicable disease, non-communicable disease, occupational safety & health, public health nutrition, the built environment, and public health in conflict, war, and peace building.

The cross-curricular subject areas of **Domain 3** are those which should be represented, in so far as possible, in all core and subject specific areas and should include: historical, cultural, sociological, economic, commercial, legal and influences on health, diversity & intersectionality, health in vulnerable groups, global public health, One Health, digital transformation in public health, health literacy, and infodemiology.

Domain 4 collates interdisciplinary professional skills of: critical thinking/evidence synthesis, knowledge translation for policy and action, communication skills, public health advocacy, negotiation, collaboration, leadership, and management.

"**One Health**" represents a great example of a cross-curricular subject area in public health training as it combines the realms of human, animal, and environmental health. This approach necessitates collaboration among diverse academic disciplines including medicine, veterinary science, ecology, sociology, and public health.

One Health prepares public health professionals to address global health challenges through a holistic and integrated approach emphasizing the importance of collaboration in solving complex, multifaceted health issues. The subject area of One Health is divided into themes and curricular elements. The infographic below represents the five themes within the CCP One Health curriculum which are presented at different educational delivery levels.

ASPHER CCP One Health Curriculum Themes

Figure 23
ASPHER CCP One Health Themes

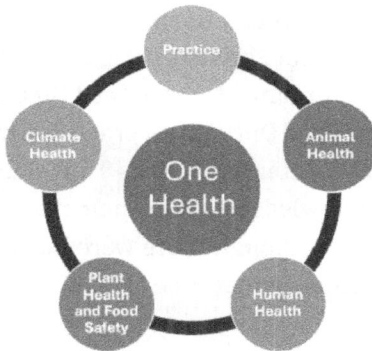

The table below presents a summary of each theme's curricular content.

Table 5	ASPHER CCP One Health themes
Practice	This theme focuses on practical interventions for public health, such as Animal Health Control through vaccination, surveillance, and culling. It includes Animal Health Surveillance for tracking diseases, Antimicrobial Resistance prevention, and the role of Animals as Environmental Sentinels. Other areas include Behavioral Sciences, Biodiversity Action, and Chemical and Physical Threats, emphasizing collaboration with regulators to address health challenges
Human Health	This theme explores the impact of food systems on human health, focusing on the Cost of Nutritious Food versus processed options, Foodborne Diseases, and Microbiome Changes from farming practices. It also examines the Human-Animal Relationship in various contexts, advocating for Sustainable Diets and addressing public trust issues in food safety.
Animal Health	This theme addresses disease management in animals, with a focus on preventing zoonotic diseases like Covid-19 and Bird Flu. It includes Farm Animal Safety and Epidemiology to ensure sustainable food systems and minimize disruptions. Control measures in animal reservoirs are highlighted to reduce the transmission of diseases from animals to humans.
Plant health and Food Safety	This theme covers plant health and food safety, focusing on Climate Disaster Impact on farming, Ecological Agriculture, and the challenges of GMOs and Plant Diseases. It promotes Natural Environmental Hygiene and biodiversity restoration in food production, ensuring safety from "Soil to Society"
Climate Health	The Climate Health theme addresses the intersection of climate change and public health, emphasizing Carbon Sequestration and the health threats of Climate Change. It covers Desertification, sustainable practices for food producers, and the impact of Heat Waves on human, animal, and plant health, with a focus on reducing Methane Emissions and protecting biodiversity.

One Health competency framework review

Core competencies for One Health are defined as "the unique competencies that all One Health professionals should have, regardless of their discipline of origin" (70). There have been several works to create such competencies. Firstly, at the Salzburg Global Seminar in 2007, participants from different types of organizations agreed that the new group of professionals called One Health practitioners should have distinct professional abilities such as "soft skills" to enhance their depth of knowledge within their respective fields of specialization, even though there were no set of core competencies established at that time (71).

Then, as cited in Frankson et al., the Bellagio Working Group was founded in Italy in 2008 to determine the core competencies necessary for leadership in global food systems (71). This was followed by a meeting that was held in Stone Mountain, Georgia on May 4-6, 2010, at which seven working groups were formed, one being training with the goal to "develop and build skills, expertise, and competencies through a One Health training curriculum and identify opportunities to integrate One Health approaches into existing curricula" (72). Afterwards, the Global One Health Core Competencies (OHCC) Working Group met several times between November 2011 and December 2012 and drafted the core competency domains (73). Last, but not least, a workshop was held in Rome on March 12-13, 2012 with the purpose of synthesizing core One Health competencies ("Rome Synthesis") (70).

Several researchers focused on core competencies for One

Health. For example, Frankson et al. noted that the afore-mentioned One Health core competency frameworks offer a shared basis for continuing professional education and training programs that go beyond concentrating on discipline-specific information and can be used as a starting point by academic institutions, governmental bodies, and regional networks to pinpoint particular core competencies pertinent to local needs, which can then be used to direct the creation of new training initiatives (71). Togami et al. recommended three core competencies for One Health education and suggested a step-by-step approach to applying One Health core competencies (74). Finally, the nine updated core competencies were provided in a paper by Laing et al. on behalf of the Network for Ecohealth and One Health (NEOH) and it was concluded that these competencies can be utilized to assess and improve existing curricula, develop new ones, or provide guidance for professional training programs for students, faculty members at universities, government employees, and frontline health professionals and policymakers at all levels (75).

Table 6 is provided below to shows the different sets of core competencies suggested by these authors.

Importance of synergistic connections: A prototype from one institution

Modern science, and arguably our understanding of the world to date, has been driven largely by the idea of reductionism. Descartes, Bacon, and others argue that to understand the whole one must first understand each of the constituent parts in isolation. This appears to be a sensible approach. Immense scientific advances have been made. However, when one examines this approach in the context of the complex world we live in, one might perhaps consider a reductionist approach as being too simplistic to truly understand the complexities of systems and to appreciate how synergism impacts on behavior and evolution. Aristotle is credited with the phrase 'The whole is greater than the sum of its parts'; Peter Corning, with the synergism hypothesis, goes further to say that the whole can not only be greater but can appear different and can achieve something that is unconnected to the individual parts. This indicates that cooperative interactions can lead to greater collective benefits. This is a key driver in the evolution of systems. It also implies that systems should seek to be synergistic to create functional advantages that would be otherwise unachievable.

Table 6	One Health core competency domains described by different researchers
Researchers	**Core Competency Domains**
Frankson et al. (71)	**Management:** Able to manage cross disciplinary teams – understands the roles and responsibilities of the team and its individual members – holds the team accountable **Communication and informatics:** Utilizes diplomacy – able to negotiate – able to resolve conflicts – achieves collaboration **Values and ethics:** Values honesty – possesses strong knowledge, and self-possesses integrity **Leadership:** Advocates for change – fosters a changed environment – understands individual and shared leadership models – possesses external awareness (social, political, legal, and cultural) **Team and collaboration:** Able to identify shared values and goals – values diversity of disciplines, culture, ideas, background, and experience, establishes trust, and thinks strategically **Systems thinking:** Awareness of the big picture and interdependence of stakeholders – understands and embraces a One Health approach is able to identify the problem and its impact on the system
Togami et al. (74)	**Health knowledge:** Demonstrated knowledge of established and evolving transdisciplinary One Health sciences, including those relevant to public health, animal health, environmental sciences, and modern agriculture **Global and local issues in humans, animals, plants and the environment:** Demonstrates an understanding of historical, cultural, political, economic, and scientific aspects of complex and emerging health problems that are amenable to the One Health approach **Professional characteristics:** Demonstrates the ability to understand and apply principles of research and evaluation methods to policy and health program implementation as well as to apply scientific findings to real-life situations
Laing et al. (75)	**Effective communication:** Engages effectively in respectful and reciprocal communication and partnerships with people from different backgrounds, disciplines, groups in society, and sectors

Figure 24
Synergistic Movement

Donella Meadow's book 'Thinking in Systems', a system is an interconnected set of elements that is coherently organized in a way that achieves something, has interconnections, and has a function of purpose (76). Systems can be simple or complex, biological or non-biological, but ultimately, they must have these fundamental components to make them into a system where synergy lies at the center. Everywhere we look we can see systems, some of which achieve an equilibrium or balance in which outputs and inputs drive forces for the advantage or disadvantage of the system but ultimately create positives or negatives that lead to balance. In contrast, complexity theory would suggest that some systems can exist in flux, where change is a constant and even small impacts can have large effects. Ultimately, whether a system is in equilibrium or is experiencing change, individual elements are engaging, influencing, and interacting with each other. So, to consider elements in isolation is, by definition, ignoring the broader picture and failing to acknowledge that we cannot think in a linear fashion but rather we must think in a circular manner whereby changes in one element 'A' can impact another element 'B' but might also impact 'A' itself because the system in which 'A' and 'B' exist is affected.

By its nature, the One Health approach aligns with systems thinking, encouraging the acceptance that humans, animals, plants, and ecosystems exist in a collective system, whereby impacts on one element affect the others, so no element should be ignored or forgotten. Importantly, One Health is not a discipline. It is not a thing that should be put alongside biology, chemistry, and physics etc. Rather, it should be treated as a fundamental belief that acts as a foundation block underpinning all disciplines. It encourages a paradigm shift in approach from reductionism to a most holistic viewpoint that emphasizes the health of animals, humans, and the environment collectively. It considers that each one is important, but each one is an element within a global system and thus is interconnected and has a collective purpose. The implication of this is that by not taking a One Health approach, such that interconnections between the environment, humans, and animals are not considered, one fails to fully consider the impacts of change within a system and one would struggle to ever fully understand one element in isolation. This means that the identification of potential consequences of decisions, the identification of the risks, or even the positives and the negatives of a decision are not complete and unintended consequences that can and will emerge leading to major impacts on the system.

The philosopher, Emmanuel Levinas emphasized our ethical obligation to the 'other' and that we are defined by how we approach this relationship. Levinas argued that when we are face-to-face with the 'other' we must recognize their vulnerabilities, similarities, and differences and respond to the 'others' needs and demands. One Health suggests that the 'other' is a human, an animal, a plant, or an ecosystem and we are obliged to consider its needs and demands alongside our own.

In recognition of this complexity and our ethical imperative, academic institutions, government agencies, industry, and the public must change the way they view the world. We must move away from reductionism, move away from isolationism, and move towards embracing a holistic One Health approach when we tackle local, national, and global challenges.

Figure 25
UCD Logo

An excellent first step to operationalize this approach is the establishment of interconnected virtual or physical centers where relationships are built and shifts in mindset are encouraged. The University College Dublin Academic Center for One Health is one such example. Built as a virtual hub within a multidisciplinary university that engages across all faculties and disciplines in an active attempt to eliminate silos and to emphasize collaboration. It focuses on education, communication, and collaboration, striving to build awareness of a One Health approach, acting as a hub, facilitating connections that take a One Health approach, and helping to build an evidence-based approach for operationalizing One Health. It works closely with organizations such as ASPHER and the quadripartite (WOAH, WHO, UNEP and FAO) as well as with members of the Irish local and national government and also builds links with industrial partners and schools through outreach programs and collaborative research.

Figure 26
UCD One Health Initiative Logo

Making a change towards a One Health approach is challenging. Our societies are not designed to work in synergistic ways that encourage connections. Governments are divided into departments. Universities are accumulations of individual disciplines that exist alongside each other where different areas of work life separate us based on how we work and what we do and even in schools we encourage streamlining, specialization, and separation. A One Health approach seeks to break down these barriers. A One Health center should aim to provide interdisciplinary education that prepares future and current professionals to work across traditional disciplinary boundaries. It should encourage a broader perspective and force people to approach problems with an alternative mindset which relies on teams rather than on individuals. Of course, this change cannot happen immediately and multiple approaches to educate the population must be taken. Current professionals or those close to graduating into the workforce can be educated with an emphasis on tools and skills that enable a One Health approach. The ability to work as part of a team, the ability to communicate, the ability to build rela-tionships, and to consider those relationships equal are skills that are translatable and actionable in almost every workplace. This goes beyond the core knowledge of a discipline and moves people into a mindset of collaboration and systems thinking.

Beyond the immediate however, an academic center should think about how to engage with future generations. It is important to build a One Health approach into everyday teaching within primary education systems, such that a One Health approach becomes a reflex that you don't have to think about. This must involve engagements with schools and with national curricula bodies. It should include building resources, lesson plans, and action packs that can be employed within schools that are problem based and encourage many people with different approaches to work together. This, if filtered and ingrained into our younger populations to encourage them when they enter the workforce, will form the future approaches that are taken, thus ultimately creating synergistic connections benefiting the system as a whole.

Figure 27
Antimicrobial Resistance

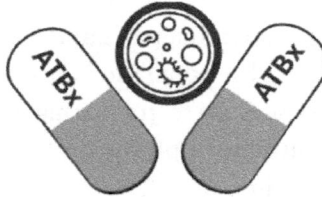

A buzzword now, post-pandemic, is 'preparedness.' During the pandemic, there was a collective movement towards protecting society. But these connections are already dissolving. An amnesia creeps in, relationships drift and siloed thinking returns. Unintended consequences continue to occur as decisions are made in isolation. But hope remains and there are concrete examples of connection building approaches that are more effective. Many countries have adopted multidisciplinary approaches to tackle antimicrobial resistance (AMR), for example. This movement acknowledges that AMR is an issue of global concern that impacts humans, animals, and the environment. Preparedness for a pandemic must take a similar approach that is broad and aggressive, recognizing the risk of not doing anything or doing things without considering a One Health perspective. Preparedness approaches must involve preparing for outbreaks scientifically, medically, and psychologically. Surveillance for potential outbreaks should be comprehensive and unbiased. The approach should emphasize the risk of no surveillance and should be independent of commercial and political interference. Improving health literacy enables the inoculation of a population with information about science and medicine that prepares a society to understand and knowledgeably discuss new outbreaks. Drugs and vaccines should be developed for what we have now and also for what might come in the future. Preparedness is broad and complex but is possible with the right approach.

By thinking of local, national, and global challenges with a One health mindset we can consider the systems involved holistically, thus enabling a collective ability to impact that challenge positively.

Conclusion

One Health pertains to a holistic systematic approach according to which humans, animals, and plants live, interact, and impact each other in any ecosystem. As such, One Health is within itself the pivot of other disciplines that necessitates an interdisciplinary approach in medicine, veterinary science, ecology, climate science, and public health. The One Health approach addresses human, animal and plant health in a collective, rather than a disconnected, discrete manner accentuating the interconnectedness of the three systems/spheres. This implies a delicate and decisive approach to the whole system. Therefore, reductionist and isolationist approaches should be abdicated and a holistic One Health approach should be adopted at local and international levels.

To implement the One Health approach, establishing educational and collaborative hubs that engage diverse stakeholders, is essential. One Health highlights the ongoing threat of antimicrobial resistance (AMR) emphasizing the need for robust food and water safety systems, understanding transmission mechanisms, and identifying both zoonotic and non-zoonotic pathogens that impact human, animal, and plant health. The rise in antimicrobial use in food-producing animals, along with the increase in greenhouse gas emissions, underscores the urgency of adopting a One Health perspective.

To address current and future challenges, ASPHER focuses on improving education and research to enhance understanding of the link between climate and health. It emphasizes the importance of integrating planetary health and One Health into public health education to better prepare the future work-force. In collaboration with GCCHE, ASPHER is working to expand the availability of climate health courses and programs globally.

Another program that has been inaugurated by ASPHER is the YPP to enhance collaborative opportunities on climate and health and to consider these topics in public health education of professionals. ASPHER COP28 statement emphasized the need for the establishment of a capable workforce for the miti-gation of the climate change impact on health. There has been a substantial endeavor in making the public health curriculum congruent across Europe to consolidate professional training programs through an exhaustive methodology with in-depth curricula on One Health being designed by experts through the ASPHER CCP.

One Health offers a groundbreaking approach to observing, analyzing, and managing health issues that intersect humans, animals, plants, and the environment. Through comprehensive risk assessment and management, this methodology enables us to identify and address potential threats across these interconnected spheres. By adopting this holistic perspective, we can proactively counteract negative outcomes, ensuring a net positive impact on health and sustainability for all.

References

1.Stephen C, Sleeman J, Nguyen N, Zimmer P, Duff J, Gavier-Widén D, et al. Proposed attributes of national wildlife health programmes. Rev Sci Tech. 2018;37(3):925-36.

2.Gu Y, Lancsar E, Ghijben P, Butler JR & Donaldson C. Attributes and weights in health care priority setting: A systematic review of what counts and to what extent. Soc Sci Med. 2015;146:41-52.

3.Traore T, Cosivi O, Dorjee S, Gongal G, Hoejskov PS, Mahrous H, et al. WHO regional One Health initiatives: Paving a transformative path towards strengthening national One Health systems and capacities in line with the Quadripartite One Health Joint Plan of Action/Initiatives regionales de l'OMS en faveur de l'approche Une seule sante: ouvrir la voie vers un renforcement des capacites et des systemes nationaux Une seule sante, conformement au Plan d'action conjoint quadripartite. Wkly Epidemiol Rec. 2023;98(48-49):645-53.

4.Brach C, Keller D, Hernandez LM, Baur C, Parker R, Dreyer B, et al. Ten attributes of health literate health care organizations. NAM Perspect. 2012:1-26.

5.Mettenleiter TC, Markotter W, Charron DF, Adisasmito WB, Almuhairi S, Behravesh CB, et al. The One Health High-Level Expert Panel (OHHLEP). One Health Outlook. 2023;5(1):18.

6.Iatridou D, Bravo A & Saunders J. One Health interdisciplinary collaboration in veterinary education establishments in Europe: mapping implementation and reflecting on promotion. J Vet Med Educ. 2021;48(4):427-40.

7.Gates MC. One Health continuing medical education: an avenue for advancing interdisciplinary communication on One Health issues. J Am Vet Med Assoc. 2009;234(11):1384-86.

8.Chakraborty S, Andrade FC & Smith RL. An interdisciplinary approach to One Health: course design, development, and delivery. J Vet Med Educ. 2022;49(5):568-74.

9.Destoumieux-Garzón D, Mavingui P, Boetsch G, Boissier J, Darriet F, Duboz P, et al. The One Health concept: 10 years old and a long road ahead. Front Vet Sci. 2018;5:14.

10.Lapinski MK, Funk JA & Moccia LT. Recommendations for the role of social science research in One Health. Soc Sci Med. 2015;129:51-60.

11.Deem SL, Lane-deGraaf KE & Rayhel EA. Introduction to One Health: An interdisciplinary approach to Planetary Health. Hoboken (NJ): Wiley-Blackwell; 2019.

12.Snowden FM. Emerging and reemerging diseases: a historical perspective. Immunol Rev. 2008;225(1):9-26.

13.Chin A, Simon GL, Anthamatten P, Kelsey KC, Crawford BR & Weaver AJ. Pandemics and the future of human-landscape interactions. Anthr. 2020;31:100256.

14.McInnes C, Kamradt-Scott A, Lee K, Roemer-Mahler A, Rushton S, Williams OD, et al. Conclusion: The transformation of global health governance. The Transformation of Global Health Governance London: Palgrave Pivot; 2014. p. 95-113.

15.Dasgupta P. Conclusion: Integrating sustainable development and health adaptation. Climate sensitive adaptation in health: Imperatives for India in a developing economy context. Berlin: Springer; 2016. p. 175-94.

16.Mesa-Lago C & Cruz-Saco MA. Conclusion: Conditioning factors, cross-country comparisons, and recommendations. Do options exist?: The reform of pension and health care systems in Latin America. Pittsburgh (PA): University of Pittsburgh Press; 2010. p. 377-428.

17.Yarcheski A, Mahon NE, Yarcheski TJ & Cannella BL. A meta-analysis of predictors of positive health practices. J Nurs Scholarsh. 2004;36(2):102-8.

18.Latif AA & Mukaratirwa S. Zoonotic origins and animal hosts of coronaviruses causing human disease pandemics: A review. Onderstepoort J Vet Res. 2020;87(1):1-9.

19.Huynh J, Li S, Yount B, Smith A, Sturges L, Olsen JC, et al. Evidence supporting a zoonotic origin of human coronavirus strain NL63. Virol J. 2012;86(23):12816-25.

20.Ryff CD & Singer B. The contours of positive human health. Psychol Inq 1998;9(1):1-28.

21.Jackson LE. The relationship of urban design to human health and condition. Landsc Urban Plan. 2003;64(4):191-200.

22.Carmona M. Public places urban spaces: The dimensions of urban design. 3rd ed. New York: Routledge; 2021.

23.Jiang B, Chang C-Y & Sullivan WC. A dose of nature: Tree cover, stress reduction, and gender differences. Landsc Urb Plan. 2014;132:26-36.

24.European Food Safety Authority and European Centre for Disease Prevention and Control. The European Union One Health 2022 zoonoses report. EFSA J. 2023;21(12):e8442.

25.Binsker U, Käsbohrer A & Hammerl JA. Global colistin use: a review of the emergence of resistant Enterobacterales and the impact on their genetic basis. FEMS Microbiol Rev. 2022;46(1):fuab049.

26.European Food Safety Authority Panel on Biological Hazards, Koutsoumanis K, Allende A, Álvarez-Ordóñez A, Bolton D, Bover-Cid S, et al. Role played by the environment in the emergence and spread of antimicrobial resistance (AMR) through the food chain. EFSA J. 2021;19(6):e06651.

27.Guenther T. Animal health. EJP NOVA Project; 2020. http://data.d4science.org/ctlg/ORIONKnowledgeHub/ab74

9ba7-e1de-41f1-a71d-98b5cfcfecce.

28.Food and Agriculture Organization of the United Nations, United Nations Environment Programme, World Organization for Animal Health, World Health Organization. One health joint plan of action (2022–2026): working together for the health of humans, animals, plants and the environment. Rome: World Health Organization; 2022.

29.World Health Organization, Food and Agriculture Organization of the United Nations & World Organisation for Animal Health. Taking a multisectoral, one health approach: a tripartite guide to addressing zoonotic diseases in countries. Food and Agriculture Organization of the United Nations, World Organization for Animal Health, World Health Organization. 2019. https://iris.who.int/bitstream/handle/10665/325620/9789241514934-eng.pdf?sequence=1.

30.Jones KE, Patel NG, Levy MA, Storeygard A, Balk D, Gittleman JL, et al. Global trends in emerging infectious diseases. Nature. 2008;451(7181):990-3.

31.Socha W, Kwasnik M, Larska M, Rola J & Rozek W. Vector-borne viral diseases as a current threat for human and animal health-One Health perspective. J Clin Med. 2022;11(11).

32.Van Boeckel TP, Brower C, Gilbert M, Grenfell BT, Levin SA, Robinson TP, et al. Global trends in antimicrobial use in food animals. Proc Natl Acad Sci USA 2015;112(18):5649-54.

33.Van Boeckel TP, Glennon EE, Chen D, Gilbert M, Robinson

TP, Grenfell BT, et al. Reducing antimicrobial use in food animals. Science. 2017;357(6358):1350-2.

34.Tiseo K, Huber L, Gilbert M, Robinson TP & Van Boeckel TP. Global trends in antimicrobial use in food animals from 2017 to 2030. Antibiotics. 2020;9(12):918.

35.Nabuurs G-J, Mrabet R, Hatab AA, Bustamante M, Clark H, Havlík P, et al. Agriculture, forestry and other land uses (AFOLU). In: P.R. Shukla, J. Skea, R. Slade, A. Al Khourdajie, R. van Diemen, D. McCollum, et al., editors. IPCC, 2022: Climate change 2022: mitigation of climate change contribution of working group III to the sixth assessment report of the Intergovernmental Panel on Climate Change. Cambridge: Cambridge University Press; 2022. p. 747-860.

36.Verkuijl C, Smit J, Green JMH, Nordquist RE, Sebo J, Hayek MN, et al. Climate change, public health, and animal welfare: towards a One Health approach to reducing animal agriculture's climate footprint. Front Anim Sci. 2024;5.

37.One Health High-Level Expert Panel, Adisasmito WB, Almuhairi S, Behravesh CB, Bilivogui P, Bukachi SA, et al. One Health: A new definition for a sustainable and healthy future. PLoS Pathog. 2022;18(6):e1010537.

38.Lerner H & Berg C. A comparison of three holistic approaches to health: One Health, EcoHealth, and Planetary Health. Front Vet Sci. 2017;4.

39.Miao L, Li H, Ding W, Lu S, Pan S, Guo X, et al. Research priorities on One Health: A bibliometric analysis. Front Public Health. 2022;10.

40.Ogunseitan OA. One Health and the environment: From conceptual framework to implementation science. Environment. 2022;64(2):11-21.

41.Essack SY. Environment: the neglected component of the One Health triad. Lancet Planet Health. 2018;2(6):e238-9.

42.Gudipati S, Zervos M & Herc E. Can the One Health approach save us from the emergence and reemergence of infectious pathogens in the era of climate change: Implications for antimicrobial resistance? Antibiotics. 2020;9(9).

43.Shafique M, Khurshid M, Muzammil S, Arshad MI, Malik IR, Rasool MH, et al. Traversed dynamics of climate change and One Health. Environ Sci Eur. 2024;36(1):135.

44.Watts N, Adger WN, Ayeb-Karlsson S, Bai Y, Byass P, Campbell-Lendrum D, et al. The Lancet Countdown: tracking progress on health and climate change. Lancet. 2017;389(10074):1151-64.

45.van Daalen KR, Tonne C, Semenza JC, Rocklöv J, Markandya A, Dasandi N, et al. The 2024 Europe report of the Lancet Countdown on health and climate change: unprecedented warming demands unprecedented action. Lancet Public Health. 2024;9(7):e495-522.

46.Mackenzie JS & Jeggo M. The one health approach—why is

it so important? Trop Med Infect Dis. 2019;4(2):88.

47.Karesh WB, Dobson A, Lloyd-Smith JO, Lubroth J, Dixon MA, Bennett M, et al. Ecology of zoonoses: natural and unnatural histories. Lancet. 2012;380(9857):1936-45.

48.Gebreyes WA, Dupouy-Camet J, Newport MJ, Oliveira CJ, Schlesinger LS, Saif YM, et al. The global one health paradigm: challenges and opportunities for tackling infectious diseases at the human, animal, and environment interface in low-resource settings. PLoS Negl Trop Dis. 2014;8(11):e3257.

49.Myers SS & Patz JA. Emerging threats to human health from global environmental change. Annu Rev Environ Resour. 2009;34(1):223-52.

50.Smith KM, Anthony SJ, Switzer WM, Epstein JH, Seimon T, Jia H, et al. Zoonotic viruses associated with illegally imported wildlife products. PLoS One. 2012;7(1):e29505.

51.Redding DW, Atkinson PM, Cunningham AA, Lo Iacono G, Moses LM, Wood JLN, et al. Impacts of environmental and socio-economic factors on emergence and epidemic potential of Ebola in Africa. Nat Commun. 2019;10(1):4531.

52.Kelly TR, Karesh WB, Johnson CK, Gilardi KVK, Anthony SJ, Goldstein T, et al. One Health proof of concept: Bringing a transdisciplinary approach to surveillance for zoonotic viruses at the human-wild animal interface. Prev Vet Med. 2017;137:112-8.

53.Daszak P, Cunningham AA & Hyatt AD. Emerging infectious diseases of wildlife - threats to biodiversity and human health. Science. 2000;287(5452):443-9.

54.Morse SS, Mazet JAK, Woolhouse M, Parrish CR, Carroll D, Karesh WB, et al. Prediction and prevention of the next pandemic zoonosis. Lancet. 2012;380(9857):1956-65.

55.Gibbs E & J. P. The evolution of One Health: a decade of progress and challenges for the future. Vet Rec. 2014;174(4):85-91.

56.Okello AL, Bardosh K, Smith J & Welburn SC. One health: past successes and future challenges in three African contexts. PLoS Negl Trop Dis. 2014;8(5):e2884.

57.World Health Organization. Component 2: Health workforce. 2024. https://www.who.int/teams/environment-climate-change-and-health/climate-change-and-health/country-support/building-climate-resilient-health-systems/health-workforce.

58.Sorensen CJ & Fried LP. Defining roles and responsibilities of the health workforce to respond to the climate crisis. JAMA Netw Open. 2024;7(3):e241435.

59.World Health Organization and Association of Schools of Public Health in the European Region. WHO-ASPHER competency framework for the public health workforce in the European region. 2020. https://iris.who.int/bitstream/handle/10665/347866/WHO-EURO-2020-3997-43756-61569-

eng.pdf?sequence=1&isAllowed=y.

60.Orhan R, Middleton J, Krafft T & Czabanowska K. Climate action at public health schools in the European region. Int J Environ Res Public Health. 2021;18(4):1518.

61.Orhan R, Middleton J & Otok R. ASPHER climate and health competencies for public health professionals in Europe. AS-PHER. 2021 https://www.aspher.org/download/882/25-10-2021-final_aspher-climate-and-health-competencies-for-public-health-professionals-in-europe.pdf.

62.Sorensen C, Campbell H, Depoux A, Finkel M, Gilden R, Hadley K, et al. Core competencies to prepare health professionals to respond to the climate crisis. PLOS Clim. 2023;2(6):e0000230.

63.Association of Schools of Public Health in the European Region. Moving towards the right to 'health for all' by training the public health and wider health workforce on climate change and health. 2021. https://www.aspher.org/download/1135/che_euhpp_statement_aspher_final-version.pdf.

64.Global Consortium on Climate and Health Education. Climate and health core concepts for health professionals [Internet]. New York: Columbia Mailman School of Public Health. 2023. https://www.publichealth.columbia.edu/file/11492/download?token=bBURLrFC.

65.Global Consortium on Climate and Health Education. European climate and health responders course. New York:

Columbia Mailman School of Public Health. 2024. https://ww w.publichealth.columbia.edu/research/programs/global-co nsortium-climate-health-education/courses/past-courses/ european-climate-health-responder-course.

66.Association of Schools of Public Health in the European Region. ASPHER Young Professionals Programme call for applicants - 2023-2024 intake. Association of Schools of Public Health in the European Region. 2023. https://www. aspher.org/news,107.html.

67.Association of Schools of Public Health in the European Region. ASPHER statement for COP28: A call for action in seven points. Association of Schools of Public Health in the European Region. 2023. https://www.aspher.org/news,111.ht ml.

68.Chambaud L, Chen T, Cadeddu C, Pinho-Gomes A-C, Ádám B, Middleton J, et al. ASPHER statement for COP28. A call for action in seven points. Public Health Rev. 2023;44.

69.Foldspang A. Provisional lists of public health core com-petencies: European Public Health Core Competencies Pro-gramme (EPHCC) for public health education. Phase 1. Brus-sels: ASPHER; 2007.

70.Hueston W, Kunkel R, Nutter F, Olson D. One health core competencies. One Health Commission. 2014. https://www .onehealthcommission.org/documents/filelibrary/library_r eferences/Hueston_Kunkel_OH_competencies_5E7BEEF4 0A553.pdf.

71.Frankson R, Hueston W, Christian K, Olson D, Lee M, Valeri L, et al. One health core competency domains. Front Pub Health. 2016;4:192.

72.Centers for Disease Control and Prevention. Operationalizing "One Health": A policy perspective - taking stock and shaping an implementation roadmap. Atlanta (GA): Centers for Disease Control and Prevention. 2011. https://www.woah.org/fileadmin/Home/eng/Media_Center/docs/pdf/meeting-overview.pdf.

73.Lee M and Global OHCC Working Group. One health core competency domains, subdomains, and competency examples, USAID RESPOND Initiative. 2013. https://dl.tufts.edu/pdfviewer/6682xh01r/9p290n711.

74.Togami E, Gardy JL, Hansen GR, Poste GH, Rizzo DM, Wilson ME, et al. Core competencies in one health education: what are we missing? Washington, DC: National Academy of Medicine; 2018. https://nam.edu/core-competencies-in-one-health-education-what-are-we-missing.

75.Laing G, Duffy E, Anderson N, Antoine-Moussiaux N, Aragrande M, Luiz Beber C, et al. Advancing One Health: updated core competencies. CABI One Health. 2023(2023):ohcs20230002.

76.Meadows DH. Thinking in systems: A primer. White River Junction (VT): Chelsea Green Publishing. 2008.

2

Sweltering Struggles: India's Fight Against the Silent Health Crisis

Abstract

India is facing a severe public health crisis as rising temperatures and intensified heat waves, fueled by climate change,
take a heavy toll on its population. Despite contributing
only seven percent to global greenhouse gas emissions, the
country is among the most vulnerable to climate-induced
adversities. Heat waves, now more frequent and severe, pose
serious health risks, especially for vulnerable populations
without access to cooling systems. Urban areas, exacerbated
by the heat island effect, face heightened challenges, while
rural regions struggle with water shortages and agricultural
impacts. By 2050, heat wave exposure is expected to rise
eightfold, threatening livelihoods, food security, and public health. Addressing this crisis requires urgent action to
strengthen public health systems, improve urban planning,
and ensure access to cooling and water resources. India must
act decisively to mitigate these risks, safeguard its population,

and build resilience in the face of a rapidly warming climate.

Keywords: Public health, mental health, heat stress, heat, climate change, temperature, climate health.

Author

Komal Mittal, Research Associate, Center for Human Progress, New Delhi, India and Youth Mentor, POP (Protect Our Planet) Movement, New York, USA

Background

India is experiencing a significant rise in temperatures, with an average increase of 0.7 degrees Celsius between 1901 and 2018, underscoring the impact of a changing climate. Ranked seventh among the list of countries most affected by climate change in 2019, India faced a unique paradox (1). While contributing only seven percent of global greenhouse gas emissions—less than the world average per capita—it remains one of the most vulnerable regions to climate-induced adversities (2). Studies show that heat waves in India are 30 times more due to climate change (3). The country is witnessing five to six heat wave events every year intensifying in frequency, duration, and severity. This escalation has transformed heat waves into a formidable public health crisis, compounded by other environmental challenges like air pollution.

Scientists estimate that temperatures in South Asia, including India, are already approximately two degrees higher than pre-industrial temperatures, with dire consequences for health, livelihoods, and ecosystems (3). In May 2022, Pakistan and India experienced severe heat waves with temperatures soaring to a staggering 51 degrees Celsius—resulting in the deaths of at least 96 people and prompted the Indian Meteorological Department to issue warnings for certain regions (3). Rising temperatures, intensified by high humidity levels in many regions of India, pose a significant threat by amplifying the adverse health effects of hot weather—commonly known as the wet bulb temperature (4)(5). Elevated humidity hinders perspiration, the body's natural cooling mechanism, causing one to feel even more intense heat than the actual recorded temperature.

Heat waves disproportionately affect vulnerable populations, particularly those without access to cooling systems and adequate healthcare. Many urban areas suffering from the urban heat island effect are facing amplified risks. These challenges highlight the need for urgent climate-adaptive urban planning and robust public health interventions.

Unprecedented and prolonged heat events are expected to occur far more frequently and cover much larger areas. Under four degrees Celsius warming, the west coast and southern India are projected to shift to new, high temperature climatic regimes threatening food security and livelihoods. With built-up urban areas rapidly becoming heat islands, urban planners will need to adopt measures to counteract this effect (6).

India's public health infrastructure is struggling to cope with an overwhelming number of heat-related illnesses particularly impacting vulnerable populations who do not have access to cooling devices (7). During heat waves, severe water shortages leave tens of thousands without essential water supplies. Every summer, reports of people dying due to scorching heat dominate the media with vast regions of the country exceeding 45 degrees Celsius. There has been no sign of respite from sweltering heat in swathes of India. In 2024, a blistering heat wave swept through large parts of India affecting health and livelihoods. The Met Office issued red warnings, stressing the danger faced by those at risk (8).

For three consecutive years, intense heat waves have severely affected various regions of India posing significant challenges to health, water resources, agriculture, power generation, and other sectors of the economy. A World Bank report warns that heat stress could lead to a loss of 34 million of the projected 80 million global job losses in India by 2030, highlighting its substantial impact on India's workforce and economic stability (8). Meanwhile, the World Health Organization (WHO), reports that heat waves caused more than 166,000 deaths worldwide between 1998 and 2017.

India is projected to experience an eightfold increase in heat wave exposure between 2021 and 2050 with a staggering 300 percent rise by the end of this century. The number of Indians exposed to heat waves increased by 200 percent from 2010 to 2016 with the central and northwestern regions being the worst affected. The eastern coast and Telangana have faced severe heat waves. Kerala recorded its first-ever heat wave in 2016, signaling the expanding reach of this crisis (9).

The number of the published articles on the subject declines as the season progresses, but this should not be mistaken as a sign of reduced importance. Instead, it reflects India's growing complacency toward extreme climatic conditions (10). Therefore, humans will have less time and scope to recover as future heat waves are projected to be more sustained and will cause higher night-time temperatures.

Future heat waves are expected to be more sustained with higher temperatures leaving limited recovery time for affected populations. This silent yet severe crisis necessitates immediate action to mitigate the health impacts of climate change. Strengthening public health infrastructure, enhancing urban planning to reduce heat island effects, and improving access to cooling systems and water resources are critical steps. India must act decisively to confront this looming threat, not only to safeguard its population's health, but also to secure its economic and environmental future. Recognizing the interplay between climate change and public health is vital to addressing the challenges posed by rising temperatures and ensuring resilience in the face of a rapidly warming world.

It's time that India faces this silent threat...

Understanding the silent health crisis

Climate change is driving an alarming rise in global temperatures, intensifying extreme weather events, particularly heat waves. The increasing frequency, duration, and intensity of heat waves is jeopardizing India's progress towards achieving the Sustainable Development Goals (SDGs) (11). Once considered a seasonal phenomenon, heat waves have emerged as a significant concern for disaster management in India due to these widespread and severe impacts on health and the environment.

Extreme weather conditions have become increasingly evident, impacting lives worldwide, particularly over the last few decades. The effects of extreme summer heat are becoming more severe, contributing to rising cases of heat-related illnesses such as heat stress and heat stroke which have significant implications for public health. According to the Indian Meteorological Department (IMD), the year 2023 saw the warmest February since 1901 with a maximum temperature of 29.5 degrees. In January 2025, Delhi recorded its hottest January in six years (12). Delhi woke up to a summer-like day with bright sunshine and minimal mist clocking a six year high maximum temperature of 26.1 degrees Celsius in January due to recurring western disturbances in the Himalayan region which led to a sharp spike in temperatures. Such temperature anomalies highlight the increasing unpredictability of India's climate (13).

Heat waves are known to be "silent killers" due to their subtle yet devastating effects on public health. Extreme heat is the leading cause of weather-related deaths and worsens pre-existing health conditions including cardiovascular disease, diabetes, mental health, and asthma (14). It also increases the risk of accidents, transmission of infectious diseases, and mortality rates. For example, India witnessed over 2,300 deaths during the 2015 heat wave, one of the deadliest globally, with states like Andhra Pradesh, Telangana, Punjab, Odisha, and Bihar bearing the brunt of the crisis. In the following year,

2016, heat continued to soar, with April seeing the highest recorded average global temperature (14). Figure 1 shows the global heat wave conditions in 2016.

Figure 1
Global heat wave conditions in 2016

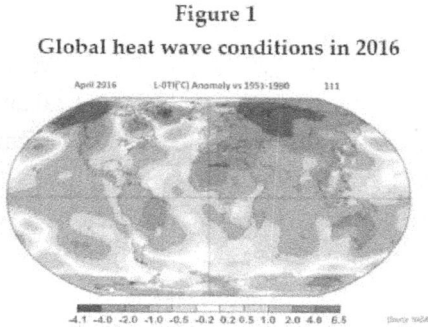

Source: Extracted from the article published by the National Institute of Disaster Management (14).

Heat stress, as a leading cause of weather-related deaths underscores the critical need for immediate action. Prolonged exposure to high temperatures, both during the day and at night, places immense strain on the human body leading to public health crises, increased mortality, and widespread socioeconomic consequences. These conditions are overwhelming healthcare systems and are impacting health service delivery capacity by disrupting health facilities, transport, and water infrastructure (15).

Heat waves disproportionately affect vulnerable populations including the elderly, infants, and people with health problems. Nearly 90 percent of India is classified as being in a heat wave danger zone with Delhi facing severe risks (16). Construction workers in particular are forced to endure grueling

67

conditions—having little choice but to work outdoors in a "hot oven" during peak summer afternoons. Though the media has now turned 'moon-eyes' about this emerging threat, it is another grim reminder about the vagaries of nature and our callous outlook.

In 2010, the Indian heat wave served as a wakeup call, underscoring the urgency of intergovernmental agency action, preparedness, and community outreach to mitigate the devastating effects of extreme heat. Since then, global and national temperature records have repeatedly been broken, reflecting the accelerating pace of planetary warming that scientists

predict is a consequence of anthropogenic emissions and poses profound long-term risk to civilization (14). Scientists warn that without immediate interventions, the situation will continue to worsen, posing profound risks to public health and sustainability.

As the frequency and severity of heat waves increase, it becomes critical to develop adaptive strategies, including reevaluating metrics for assessing climate vulnerability and implementing targeted mitigation efforts. The question that arises is "How hot is too hot?" This demands urgent attention as India grapples with the growing toll of heat-related illnesses and fatalities (14).

The silent health crisis triggered by heat waves is not merely a consequence of climate change but a stark reminder of the vulnerabilities within our social and economic systems. It is also important to understand the plight of vulnerable populations who bear a disproportionate burden of this escalating crisis (16).

Vulnerable populations

India's fight against climate change reveals that certain groups bear the brunt of the escalating heat crisis. Vulnerable populations—including low-income communities, the elderly, children, women, and outdoor workers—face disproportionate impacts due to limited resources, preexisting inequalities, and inadequate infrastructure.

Research has shown a clear relationship between extreme heat conditions and rising mortality rates, emphasizing the serious risk posed to vulnerable populations (17). Heat waves place a significant burden on low-income households which often lack access to adequate water, cooling facilities, and resources to cope with extreme temperatures. This challenge is further compounded by the urban heat island effect where densely built-up urban areas, especially underprivileged neighborhoods, experience higher temperatures than their rural surroundings. Poor neighborhoods are disproportionately affected with residents often living in inadequately ventilated homes that lack effective cooling systems. Due to poor urban planning and sub-optimal housing design low-income communities suffer from thermal discomfort, higher energy consumption, and economic distress. Residents often resort to borrowing large sums to purchase air conditioners contributing to growing heat poverty (18)(19)(20)(21).

Poor urban planning contributes to thermal discomfort in low-income settlements. Housing in these areas often lacks proper spatial design and ventilation which leads to overheating. Poor and marginalized communities usually live clustered in ghettos which have poorly ventilated homes with limited access to cooling methods (22).As heat waves become more frequent and intense, India's most impoverished people are being pushed into heat poverty as they borrow large sums to buy air conditioners for their homes. The India air conditioning market is expected to grow to US$5 billion by 2028 (23). This demand is increasingly financed by loans, a trend that has prompted concerns in the Reserve Bank of India (24)(25). Despite this, local governments frequently prioritize immediate housing needs over creating thermally comfortable environments.

Additionally, heat waves strain essential resources such as water and electricity, disproportionately affecting low-income groups. These households are also more likely to face food shortages and income loss due to the economic consequences of extreme heat.

Many populations in tropical and subtropical climates face constant exposure to high temperatures whereas in mid to high latitudes, excessive heat is seasonal. However, research on the long-term effects of chronic heat exposure and humidity on individuals is very limited.

Heat-related mortality among people over 65 years of age increased by approximately 85 percent between 2000-2004 and 2017-2021 (26). Population aging and the growing prevalence of non-communicable diseases such as respiratory and cardiovascular diseases, diabetes, dementia, renal disease, and musculoskeletal disease, makes older adults more vulnerable to heat waves. Prolonged exposure to high temperatures can exacerbate existing heat conditions and lead to heat stress and heat stroke (27).

As India grapples with increasingly intense heat waves, children, one of its most vulnerable populations, face significant health and developmental risks. Prolonged exposure to high temperatures is more than just an issue of discomfort, it can lead to severe health complications. Children face significant risks from extreme heat due to their higher body surface area relative to their weight and slower acclimatization to temperature changes. Increasing frequency and intensity of heat waves in India is alarming and demands immediate action (28). Prolonged high temperatures can lead to long-term health issues affecting children's physical health and cognitive development (28). These extreme heat conditions increase the likelihood of dehydration, heat exhaustion, and heat stroke. A lack of cooling facilities in schools amplifies this vulnerability, leading to absenteeism and reduced academic performance further widening socioeconomic disparities in education (22). Save the Children has emphasized the urgent need to address the escalating impact of extreme weather conditions exacerbated by the global climate crisis (28).

Women in vulnerable communities face unique challenges related to extreme heat. According to an article by BBC in June 2024, "It is not difficult to fall ill in Delhi. Life is tough. A third of residents live in substandard and congested housing. The city's 6,400 slums—home to more than a million—lack adequate cooling and face seasonal livelihood crisis. Men fall ill working outdoors; women fall sick after spending extended periods in kitchen settings with traditional stoves (29)." The compounded effects of social inequities further exacerbate their vulnerability to climate change.

Occupational exposure significantly increases vulnerability in agriculture workers who perform outdoor labor under intense heat and have prolonged exposure to high temperatures. Predominantly from marginalized communities, these workers lack access to protective measures, proper hydration, and adequate rest periods which heightens their risk to heat stress and related health issues. The economic strain on these workers due to reduced productivity during heat waves increases socioeconomic challenges (22).

The health impacts of extreme heat are more severe in urban areas where residents experience higher and sustained temperatures compared to surrounding areas. While urban areas experience the urban heat island effect, rural communities are not immune to facing significant challenges of extreme heat. This UHI effect is caused by a combination of heat absorbing surfaces, the trapping of hot air between buildings, limited tree cover, other heat trapping, and heat-inducing factors such as fuel combustion and air conditioning resulting in average temperatures being hotter than in the surrounding areas. The disparity in resources between rural and urban regions often leaves rural populations more vulnerable to the long-term effects of heat waves (14).

Extreme heat is not only causing severe health problems, it

is also exacerbating the poverty gap in India. Rising temperatures are projected to cost India 2.8 percent of its GDP by 2050 as per a 2018 World Bank report (30). And a 2022-2923 Reserve Bank of India report suggests that, by 2100, without mitigation efforts, this could rise to anywhere from three percent to ten percent (31). During India's record-breaking heat wave in March and June 2023, temperatures soared to 52.3 degrees Celsius leading to over 40,000 cases of heat stroke and more than 100 deaths (32).

Looking ahead, long-term projections for 2050 indicate that heat waves in India could surpass the survivability threshold for a healthy person resting in the shade. Moreover, they will impact labor productivity, economic growth, and quality of life of around 310-480 million people (33). According to the Ministry of Home Affairs, an estimated 730 people in India died from heat stroke in 2022—almost double the number of those who died in 2021 (34)(35). The economic toll of extreme heat is multifaceted. Reduced labor productivity, increased healthcare expenses, and the rising cost of cooling solutions strain household and the national economy. Additionally, rising temperatures threaten to widen the poverty gap disproportionately impacting the most vulnerable.

The Intergovernmental Panel on Climate Change (IPCC) has warned that the frequency and severity of extreme weather events including heat waves will continue to rise unless immediate and concerned global action is taken (36). Addressing the disproportionate impacts on vulnerable groups requires a multi-faceted approach: 1) develop and enforce heat action plans that prioritize vulnerable populations including low-income groups, the elderly, children, and outdoor workers; 2) promote climate-resilient urban planning and thermally efficient housing; 3) increase access to affordable and sustainable cooling technologies in vulnerable communities; 4) educate communities about heat-related health risks and protective measures; and 5) encourage international collaboration to address the root causes of climate change and reduce greenhouse gas emissions.

The state of India's healthcare system

India's healthcare system faces significant challenges in addressing the escalating health crisis induced by climate change. One of the most pressing issues is the disparity in healthcare infrastructure between urban and rural areas. Rural communities, which often lack adequate healthcare facilities, face severe barriers in accessing timely medical care during climate-induced health emergencies like heat waves. On the other hand, urban hospitals are already overburdened with high patient loads and limited resources, making it difficult to accommodate the surge in demand caused by extreme weather events.

The effects of heat waves, exacerbated by rapid urbanization, population growth, and the urban heat island effect, have placed immense strain on India's healthcare system (37)(38). Extreme heat events resulting in increased ambulance callouts, overcrowded emergency departments, higher hospitalization rates, and shortages of medicines, inflating healthcare expenditure. For instance, studies show that a temperature increase of just one degree Celsius above a city-specific threshold can lead to a significant rise in mortality and hospital admissions (39).

India's healthcare system is not adequately prepared to handle the health impacts of climate change. The lack of climate adaptation measures and inadequate disaster preparedness amplify the risks posed by heat waves and other extreme events. Hospitals often fail to incorporate climate-specific risks into their hazard vulnerability analyses, leaving them ill-equipped to manage surges in patient demand during emergencies (40).

The International Panel on Climate Change projects that global warming of 1.5 degrees Celsius could be reached by the 2030s if mitigation efforts are not intensified (41)(42). This warming is expected to worsen the frequency and intensity of extreme heat events, making it imperative for healthcare systems to prioritize climate resilience. However, many healthcare facilities still lack the infrastructure and protocols needed to effectively respond to these challenges by using triage systems, updated emergency plans, and climate-specific training for healthcare workers (43)(44).

Climate change disproportionately impacts marginalized populations including the scheduled castes (SCs), scheduled tribes (STs), women, the elderly, and low-income groups, who lack access to essential healthcare services (22). About 85 percent of India's workforce is employed in the informal sector where workers endure extreme conditions without adequate safety measures or social protections. This vulnerability is compounded by limited access to real-time weather updates and heat wave warnings, leaving rural and underserved populations more exposed to heat-related illnesses (22).

Disparities in healthcare access exacerbate the immediate and long-term health impacts of heat waves undermining income-generating capabilities and perpetuating cycles of poverty (22). For example, delayed treatment for heat-related illnesses can lead to severe health outcomes, particularly among those unable to afford medical care or take time off work.

The economic burden of climate-induced health crises is staggering. A study by the Asian Development Bank highlights how inadequate healthcare access during heat waves worsens health outcomes for marginalized groups, while health-related costs, including hospital admissions and emergency visits, contribute to significant financial losses (22). In the U.S., for instance, the health-related costs of 10 climate events in 2012 totaled $10 billion, providing a cautionary example for the potential financial strain on India's healthcare system (45).

Healthcare professionals also face increased physical and mental health risks due to the demands of managing climate-induced crises. As the frequency of extreme weather events increases, the risk of burnout among medical staff rises, emphasizing the need for a larger, climate-ready workforce (46)(47).

To address these challenges, India's healthcare system must adopt a holistic approach. Hospitals should integrate climate-specific risks into their hazard vulnerability analyses, update emergency plans, and invest in climate-resilient infrastructure such as heat-resistant buildings and energy-efficient cooling systems (48). Special attention should be given to identifying and supporting vulnerable populations in the surrounding communities.

The healthcare sector also has a crucial role in mitigating its own climate impact. Globally, healthcare accounts for approximately five percent of carbon emissions (49). By adopting sustainable practices and reducing its carbon footprint, India's healthcare system can lead by example in combating climate change. This includes transitioning to renewable energy, optimizing supply chains, and promoting green healthcare facilities.

As heat waves and other climate-related events intensify, healthcare systems must be prepared to manage the growing burden of climate-sensitive health outcomes. This requires comprehensive policy measures that address the vulnerabilities of informal workers and marginalized groups, improve access to real-time information, and strengthen disaster preparedness.

The World Health Organization's framework for climate-resilient and low-carbon health systems offers valuable guidance in this regard. By incorporating sustainable financing, robust governance, community engagement, and capacity-building initiatives, India can create a healthcare system capable of adapting to the dual challenges of climate change and public health.

The journey to resilience is urgent and complex but by prioritizing equity, innovation, and sustainability, India's healthcare system can safeguard the wellbeing of its people and ensure a healthier, more secure future.

Government and policy interventions

India has established frameworks to address the intersection of climate and health such as the National Action Plan on Climate Change (NAPCC). This comprehensive policy initiative comprises various missions targeting renewable energy, water conservation, sustainable agriculture, and energy efficiency, indirectly addressing the health impacts of climate change. At the state and local levels, governments have implemented Heat Action Plans (HAPs) aimed at mitigating the adverse effects of extreme heat events. For instance, the National Disaster

Management Authority (NDMA) actively supports states in drafting and implementing HAPs, particularly in vulnerable regions like Odisha, Maharashtra, and Bihar (50).

However, despite these efforts, significant gaps remain in policy implementation and enforcement. Challenges include inadequate funding, limited coordination among governmental departments, and a lack of awareness about the importance of climate resilience in public health. Bridging these gaps will require stronger governance structures, robust inter-agency collaboration, and increased investment in capacity-building initiatives.

Grassroots and community-level actions play a crucial role in addressing climate-induced health risks. Local organizations and NGOs have been instrumental in creating awareness and driving action, particularly in rural and underserved areas. For example, the Ahmedabad Municipal Corporation's initiative to implement cool rooftops at Shardaben Hospital effectively reduced heat stress and was later extended to all hospitals under its jurisdiction as part of its heat safety plan (51). Such initiatives demonstrate the transformative impact of community-led efforts when aligned with broader governmental strategies.

Public awareness campaigns, targeting vulnerable populations, can however empower communities to adopt heat-resilient practices. By educating individuals about heat-related illnesses and preventive measures, NGOs and local organizations can help mitigate the health impacts of extreme heat events. Community cooling centers, heat-resistant housing, and access to clean water are essential components of these grassroots efforts (49).

Technological advancements and innovative solutions offer promising avenues for mitigating the health impacts of climate change. Renewable energy technologies and green urban planning can significantly reduce urban heat island effects and air pollution. For instance, integrating green roofs, reflective building materials, and urban forests into city planning can enhance natural cooling mechanisms and improve air quality.

Advances in health monitoring systems also hold great potential. By leveraging data analytics, artificial intelligence, and geospatial mapping, authorities can predict and mitigate climate-related health crises more effectively. For example, AI-driven models can analyze historical weather data and real-time conditions to provide early warnings about heat waves and other extreme weather events. This predictive capability enables healthcare systems to prepare adequately and allocate resources to vulnerable regions.

Additionally, innovative cooling systems, such as low-cost, low-energy cooling technologies and reflective paints, are critical for reducing heat exposure particularly in resource-constrained settings. Experimentation with heat-resistant crops and green building techniques further demonstrates the potential of innovation in addressing climate challenges.

The health system's preparedness for the growing threat of heat waves must be rooted in resilience, a concept defined by the IPCC as the capacity of social, economic, and environmental systems to adapt, respond, and reorganize while maintaining their essential functions. Hospitals and healthcare facilities must prioritize infrastructure resilience,

including investments in green and cool rooftops to reduce their contribution to urban heat islands and lower energy demands (49).

The healthcare sector, which contributes approximately five percent of global carbon emissions, must lead by example in adopting sustainable practices (49). International efforts, such as the WHO's Operational Framework for Climate-Resilient and Low-Carbon Health Systems, provide a roadmap for achieving dual goals— building resilience to climate shocks and reducing the sector's carbon footprint (52). Indian healthcare leaders can draw inspiration from initiatives like England's National Health Service collaboration with WHO on decarbonization strategies (49).

A resilient health system also requires enhanced service delivery, governance, and sustainable financing. Training healthcare professionals to address climate-induced health risks and fostering community engagement are critical components of this strategy. By integrating climate risks into health policies and planning, India can strengthen its healthcare system's ability to adapt to the growing challenges posed by climate change.

Steps forward

As India faces escalating climate challenges, the increasing frequency and intensity of heat waves and extreme heat events, pose severe threats to human health, particularly to vulnerable populations. To mitigate these challenges, a concerted and multisectoral approach is imperative to build resilience and protect vulnerable populations. Public-private partnerships can drive innovative solutions to reduce heat impacts such as: 1) developing low-cost technologies, including energy-efficient cooling systems and reflective paints; 2) scaling-up green building techniques for better thermal regulation; and 3) leveraging technologies like drone mapping to identify urban heat islands and inform localized interventions.

Collaboration between different ministries, departments, and non-governmental organizations is vital to align national, state, and local efforts under a unified framework. Public awareness is critical in addressing the health impacts of heat waves. Campaigns should focus on educating citizens about heat-related health risks and preventive measures, training healthcare workers and local leaders to recognize and respond to heat illnesses effectively, and promoting sustainable practices, such as tree planting, urban greening, and efficient water management.

Targeted communication efforts must ensure inclusive outreach to marginalized groups including women, children, the elderly, and outdoor laborers. The National Disaster Management Authority (NDMA)'s public awareness initiatives and Heat Action Plans are steps in the right direction but require broader implementation.

To effectively address the health risks posed by climate-induced heat challenges, India must adopt a long-term vision rooted in sustainable practices, improved healthcare access, and robust policy interventions. One critical area is sustainable urban planning. Cities should be redesigned to mitigate the accumulation and generation of urban heat. This can be achieved through expanding green spaces, utilizing heat-reflective building materials, and

incorporating tree-lined streets in urban layouts. These measures can lower ambient temperatures, improve air quality, and provide natural cooling mechanisms.

Healthcare system strengthening is another vital pillar of resilience. Hospitals and healthcare facilities must be equipped with resources to address heat-related illnesses especially in rural and underserved regions where vulnerabilities are pronounced. This includes the installation of cooling systems, provision of adequate water supplies, and training healthcare professionals to recognize and manage heat-related illnesses. Additionally, enhancing emergency response mechanisms during heat waves is essential for minimizing fatalities and long-term health consequences.

Policy reforms and financial investment play a pivotal role in building climate resilience. Special attention must be given to states like Odisha, Maharashtra, and Bihar, which are among the most heat-vulnerable regions in India. Policies should address the unique challenges faced by marginalized groups including women, children, the elderly, and persons with disabilities, ensuring that they are adequately protected from heat risks. These policies must also integrate gender-sensitive and socially inclusive frameworks acknowledging that certain population groups bear a disproportionate burden during extreme heat events.

Aligning these measures with global frameworks, such as the WHO's health system resilience framework, will further strengthen India's efforts. This framework emphasizes the need to integrate climate risks into health system planning, enhance community engagement, and foster leadership to manage uncertainties effectively. By embedding resilience into healthcare delivery and infrastructure development, India can create systems capable of withstanding the growing challenges of climate change.

Health must be positioned at the forefront of India's climate action strategies. Enhancing early warning systems for heat waves is a critical component of this prioritization. Mapping vulnerable populations such as those living in slums, outdoor workers, and individuals with preexisting health conditions should be an integral part of the strategy. By identifying and targeting these groups, authorities can provide tailored interventions including counseling, early warnings, and access to cooling centers.

Investing in heat-resilient infrastructure is equally crucial. Schools, community centers, and healthcare facilities must be upgraded or newly built to withstand extreme weather conditions including high temperatures. Heat-resistant materials, natural ventilation designs, and rooftop solar cooling systems can ensure that these spaces remain functional and comfortable during heat waves. Such infrastructure is particularly important in rural areas where access to healthcare and other essential services is limited.

Occupational safety programs must also integrate climate change considerations. Workers in sectors like agriculture, construction, and transportation are particularly vulnerable to extreme heat. These programs should include measures to reduce exposure, provide shaded areas, and offer hydration facilities. Addressing mental health challenges associated with prolonged exposure to heat and climate-related stresses is equally important. Finally, bridging

socioeconomic gaps through financial assistance, skill development, and livelihood support will further enhance resilience among marginalized communities.

India cannot address the impacts of climate change in isolation. Global cooperation is essential to ensure effective mitigation and adaptation. Stronger international commitments are needed to reduce greenhouse gas emissions and combat the root causes of global warming. Developing countries like India require financial and technical support from the international community to build climate-resilient infrastructure and implement large-scale adaptation strategies.

Advocacy for global action must include pushing for commitments to support technology transfers, facilitate innovation, and finance climate-resilient projects in vulnerable regions. India must also play an active role in global forums urging developed nations to fulfill their climate finance pledges and adopt ambitious emission-reduction targets. By fostering international collaboration, India can strengthen its own climate resilience while contributing to global efforts to combat the climate crisis.

Concluding comments

The growing frequency and severity of heat waves and extreme heat events are stark reminders of the increasing climate crisis. These events disproportionately impact marginalized communities but their consequences extend to all levels of society. Addressing this challenge requires urgent, coordinated, and inclusive action at both the local and national levels.

India's vision of achieving zero heat-related mortality is

ambitious but achievable through a dual approach. On one hand, top-down strategies—such as policy reforms, large-scale infrastructure investments, and integrating health into climate strategies will provide the structural backbone for resilience. On the other hand, bottom-up measures including community-driven initiatives, public awareness campaigns, and localized interventions will empower citizens to actively participate in reducing heat risks.

Ultimately, tackling the health impacts of heat waves is a collective responsibility. Governments, the private sector, and civil society must work together to protect vulnerable populations, prioritize health in climate action, and create sustainable urban and rural environments. Actions taken today will shape a future that is not only resilient to climate challenges but also equitable, inclusive, and protective of generations to come. This is the time for bold decisive action to address the sweltering struggles of our time and ensure a safer, healthier, and more sustainable India.

References

1.Eckstein D, Vera K, & Laura S. Global Climate Risk Index 2021: Who suffers most from extreme weather events? Weather-related loss events in 2019 and 2000-2019. Germanwatch. 2021 Jan. https://www.germanwatch.org/sites/default/files/Global%20Climate%20Risk%20Index%202021_2.pdf

2.Environment UN. Emissions gap report 2019. United Nations Environment Programme. 2019 Nov 26. https://www.unep.org/resources/emissions-gap-report-2019

3.Anderson K. Why is climate change a huge challenge for India. Leaf by Greenly. 2023 Jun 21. https://greenly.earth/en-gb/blo g/ecology-news/why-is-climate-change-a-huge-challenge-for-india

4.Soundarajan V. Is India ready to face the silent disaster? Down To Earth. 2018. https://www.downtoearth.org.in/clima te-change/is-india-ready-to-face-the-silent-disaster—60 962

5.Anderson K. Why should we care about the wet bulb temper-ature? Leaf by Greenly. 2024 Sep 26. https://greenly.earth/en -us/blog/ecology-news/why-should-we-care-about-the-we t-bulb-temperature

6.World Bank. India: Climate change impacts. World Bank Group. 2013 Jun 19. https://www.worldbank.org/en/news/fea ture/2013/06/19/india-climate-change-impacts

7.Kiruba CR. Sweltering heat turns public hospitals into poten-tial 'death traps'. The Hindu. 2024 Aug 23. https://www.thehi ndu.com/sci-tech/health/sweltering-public-hospitals-turn-into-death-traps-for-poor-communities/article68551498.e ce

8.PTI. No sign of respite from sweltering heat in swathes of India. India News: The Indian Express. 2024 May 21. https://i ndianexpress.com/article/india/no-sign-respite-sweltering-heat-swathes-india-9343508/

9.Alexander U. Ecological myths, warming climates and the

end of nature. The Caravan. 2019 Jun 01. https://caravanmag azine.in/environment/ecological-myths-warming-climates-end-nature

10.Garg S. Sweltering struggles: Navigating India's heatwave health risks. Society for the Study of Peace and Conflict. 2024 Jul 23. https://pure.jgu.edu.in/id/eprint/8188/1/Sweltering%2 0Struggles_%20Navigating%20India%27s%20Heatwave%2 0Health%20Risks%20_%20Society%20for%20the%20Stud y%20of%20Peace%20and%20Conflict.pdf

11.Padma TV. With heat waves projected to worsen across India, do vulnerability assessments and heat action plans suffice? Mongabay. 2023 May 23. https://india.mongabay.com/2023/ 05/as-heatwaves-projected-to-worsen-across-india-do-vu lnerability-assessments-and-heat-action-plans-suffice/

12.Siddiqui U & Fareed R. 'In a hot oven': India heatwaves take a toll on most vulnerable. Climate Crisis News. Al Jazeera. 2023 May 30. https://www.aljazeera.com/news/2023/5/30/in-a-h ot-oven-india-heatwaves-take-a-toll-on-most-vulnerable

13.Gandhiok J. Delhi logs warmest January day in six years. Hindustan Times. 2025 Jan 20. https://www.hindustantimes. com/cities/delhi-news/delhi-logs-warmest-january-day-in -six-years-101737311752778.html

14.Gupta A & Guleria S. Heat wave 2016, India: A documenta-tion study (based on State of Telangana and Odisha status). 2017 Aug 31. https://nidm.gov.in/PDF/pubs/heat_wave_18.p df

15.World Health Organization. Heat and health. World Health Organization. 2024 May 28. https://www.who.int/news-roo m/fact-sheets/detail/climate-change-heat-and-health

16. The Hindu Bureau. 90% of India vulnerable to heatwave impact, says study. The Hindu. 2023 Apr 20. https://www.the hindu.com/news/national/90-of-india-vulnerable-to-heat wave-impact-says-study/article66756784.ece

17.Berger T, Chundeli FA, Pandey RU, Jain M, Tarafdar AK & Ramamurthy A. Low-income residents' strategies to cope with urban heat. Land Use Policy. 2022 Aug 01; 119:106192. https://www.sciencedirect.com/science/article/pii/S0264837 722002198

18.Ali SB & Patnaik S. Thermal comfort in urban open spaces: Objective assessment and subjective perception study in trop-ical city of Bhopal, India. Urban Climate. 2018 Jun; 954-967 (24). https://www.scopus.com/record/display.uri?eid=2-s2.0 -85039073445&origin=inward&txGid=46482c452cd9a1086cf 66c609af51b20

19.Chen C-f, Xu X, & Day JK. Thermal comfort or money saving? Exploring intentions to conserve energy among low-income households in the United States Science Direct. Energy Research & Social Science. 2017 Apr; 61-71 (26). https://www. sciencedirect.com/science/article/abs/pii/S221462961730009 9

20.Nazarian N, Acero JA & Norford L. Outdoor thermal comfort autonomy: Performance metrics for climate-conscious urban

design. Build Environ. 2019 May 15; 155:145–60. https://www
.sciencedirect.com/science/article/pii/S036013231930188X

21. Wang J, Kuffer M, Sliuzas R & Kohli D. The exposure of
slums to high temperature: Morphology-based local scale
thermal patterns. Science of the Total Environment. 2019 Feb
10; 650:1805–17. https://www.sciencedirect.com/science/arti
cle/pii/S0048969718337811

22. Akhter S & Chauhan SV. Divided by class: How heat waves
expose socio-economic fault lines. Peoples Democracy.https:
//peoplesdemocracy.in/2024/0616_pd/divided-class-how-h
eat-waves-expose-socio-economic-fault-lines

23. Business Standard. India's room AC market likely to touch
$5 billion by FY28: Voltas. Business Standard https://www.bu
siness-standard.com/industry/news/india-s-room-ac-mark
et-likely-to-touch-5-billion-by-fy28-voltas-12307020049
8_1.html

24. Ananda J. Over 1/4th of room ACs bought via consumer
credit schemes, EMIs. The Indian Express. 2020 Feb 09.
https://www.newindianexpress.com/business/2020/Feb/0
9/over-14th-of-room-acs-bought-via-consumer-credit-sc
hemes-emis-2100885.html

25. Gopakumar G. "Worried over high growth in personal
loans." Mint. 2023 Oct 07. https://www.livemint.com/news/i
ndia/worried-over-high-growth-in-personal-loans-116966
15302022.html

26. Lancet. Explore our data. Lancet Countdown. https://lanc etcountdown.org/explore-our-data/

27. World Health Organization. Heat and health. World Health Organization. 2024 May 28. https://www.who.int/news-roo m/fact-sheets/detail/climate-change-heat-and-health

28. Kamal N. Heatwaves in India: Silent crisis impacting our children. Times of India. 2024 May 31. https://timesofindia.i ndiatimes.com/india/heatwaves-in-india-silent-crisis-imp acting-our-children/articleshow/110598821.cms

29. Biswas S. India heat: Inside Delhi's first emergency room to tackle crisis. BBC. 2024 Jun 20. https://www.bbc.com/news /articles/cn00nkzdvkjo

30. Muthukumara M, Sushenjit B, Shun C, Anil M & Thomas M. Publication: South Asia's hotspots: Impacts of temperature and precipitation changes on living standards. Open Knowl-edge Repository. 2018 Jun 28. https://openknowledge.worldb ank.org/entities/publication/8a6180ce-4e60-5ec2-b869-74 cfaad0dbb1

31. Reserve Bank of India. Report on currency and finance. Reserve Bank of India. 2024 Jun 29. https://rbi.org.in/Scripts/ AnnualPublications.aspx?head=Report%20on%20Currency% 20and%20Finance

32. Economic Times. Summer of '24: India's heatwave turns deadly with over 100 deaths, 40,000 suspected heatstroke cases. The Economic Times. 2024 Jun 20. https://economicti

mes.indiatimes.com/news/india/summer-of-24-indias-hea
twave-turns-deadly-with-over-100-deaths-40000-suspect
ed-heatstroke-cases/articleshow/111127851.cms?from=mdr

33. Kapoor C. 'Heat poverty': A growing threat in India. Global
Health NOW. 2024 Jul 16. https://globalhealthnow.org/2024-
07/heat-poverty-growing-threat-india

34.National Crime Records Bureau. Accidental deaths suicides
in India 2022. National Crime Records Bureau, Ministry of
home Affairs. https://www.ncrb.gov.in/uploads/files/Acciden
talDeathsSuicidesinIndia2022v2.pdf

35. Mohanasundaram M, Abhiyant T, Vijay L, Andreas M,
Singh AK, Upasona G, Debkumar P, & Chandrakant L. Heat
stress in India: A review. Preventive Medicine: Research &
Reviews. 2024 May-Jun. https://journals.lww.com/pmrr/fullt
ext/2024/01030/heat_stress_in_india___a_review.6.aspx

36. The CSR Journal. Heatwaves in India: The silent crisis
impacting our children. The CSR Journal. 2024 May 31.
https://thecsrjournal.in/corporate-social-responsibility-csr-
news-heatwaves-in-india-the-silent-crisis-impacting-our-
children/

37. Kong J, Zhao Y, Carmeliet J & Lei C. Urban heat island and its
interaction with heatwaves: A review of studies on mesoscale.
Sustainability. 2021 Jan;13(19):10923. https://www.mdpi.com
/2071-1050/13/19/10923

38. Ritchie H, Samborska V, & Roser M. Urbanization. Our

World in Data. 2024 Dec. https://ourworldindata.org/urbaniz
ation

39. Son Ji-Y, Bell ML, Lee J-T. The impact of heat, cold, and
heat waves on hospital admissions in eight cities in Korea.
International Journal of Biometeorology. 2014 Nov; 58(9):
1893-903. https://pubmed.ncbi.nlm.nih.gov/24445484/

40. WOTR. Scorching summers: Understanding heat waves,
heat stress, and their impact on rural India. WOTR: Rejuve-
nating Communities & Ecosystems. https://wotr.org/blog/sc
orching-summers-understanding-heat-waves-heat-stress-
and-their-impact-on-rural-india/

41. Working Group II Contribution to the Fourth Assessment
Report of the Intergovernmental Panel on Climate Change.
Climate change 2007: Impacts, adaptation, and vulnerability.
Intergovernmental Panel on Climate Change 2007. https://w
ww.ipcc.ch/site/assets/uploads/2018/03/ar4_wg2_full_repo
rt.pdf

42. The Paris Agreement. What is the Paris Agreement. United
Nations Climate Change. https://unfccc.int/process-and-me
etings/the-paris-agreement

43. Rizmie D, Preux L, Miraldo M, & Atun R. Impact of extreme
temperatures on emergency hospital admissions by age and
socio-economic deprivation in England. Social Science &
Medicine. 2022 Sep. https://www.sciencedirect.com/scien
ce/article/pii/S0277953622004993?via%3Dihub

44. Schmeltz MT,Petkova EP, & Gamble JL. Economic burden of hospitalizations for heat-related illnesses in the United States, 2001–2010. Int. J. Environ. Res. Public Health. 2016; 13(9), 894. https://www.mdpi.com/1660-4601/13/9/894

45. Limayr VS, Max W, Constible J, & Knowlton K. Estimating the health-related costs of 10 climate-sensitive U.S. events during 2012. AGU Advancing Earth and Space Sciences. 2019 Sep 17. https://agupubs.onlinelibrary.wiley.com/doi/10.1029/2019GH000202

46. Johnson H, Kovats RS, McGregor G, Stedman J, Gibbs M, Walton H, Cook L, & Black E. The impact of the 2003 heat wave on mortality and hospital admissions in England. Health Stat Q. 2005 Spring; (25):6-11. https://pubmed.ncbi.nlm.nih.gov/15804164/

47. Schaffer A, Muscatello D, Broome R, Corbett S, & Smith W. Emergency department visits, ambulance calls, and mortality associated with an exceptional heat wave in Sydney, Australia, 2011: A time-series analysis. Environmental Health. 2012 Jan 24. https://ehjournal.biomedcentral.com/articles/10.1186/1476-069X-11-3

48. Scheelbeek PFD, Dangour AD, Jarmul S, Turner G, Sietsma AJ, Minx JC, et al. The effects on public health of climate change adaptation responses: A systematic review of evidence from low- and middle-income countries. Environ Res Lett. 2021 Jul; 16(7):073001. https://www.ncbi.nlm.nih.gov/pmc/articles/PMC8276060/

49. Patel L, Conlon KC, Sorensen C, McEachin S, Nadeau K, Kakkad K, et al. Climate change and extreme heat events: how health systems should prepare. NEJM Catal. 2022 Jun 15.;3(7): CAT.21.0454. https://catalyst.nejm.org/doi/full/10.1056/CAT. 21.0454

50. World Bank. South Asia's hotspots: Impacts of temperature and precipitation changes on living standards: Report preview. World Bank Group. https://documents.worldbank.or g/en/publication/documents-reports/documentdetail/en/26 4561510162968505

51. Singh C, Madhavan M, Arvind J & Bazaz A. Climate change adaptation in Indian cities: A review of existing actions and spaces for triple wins. Urban Climate. 2021 Mar 01; 36:100783. https://www.sciencedirect.com/science/article/pii/S2212095 521000134

52. Ansah EW, Amoadu M, Obeng P & Sarfo JO. Health systems response to climate change adaptation: A scoping review of global evidence. BMC Public Health. 2024 Jul 29.; 24:2015. https://www.ncbi.nlm.nih.gov/pmc/articles/PMC11285469/

3

Transgenics: Balance Innovation, Ethics, and Sustainability in a Changing World

Abstract

Nowadays transgenic organisms play a key role in human well-being and food security and this positive impact is expected to increase in the years to come. Historically, the human species' survival has depended in their relationship with others, from wolves to rice or bacteria, our survival is the result of the domestication process which was first attained by artificial selection and now through a more specialized method, genetic transformation.

Genetic transformation has revolutionized the world as we know it, tackling issues such as human health and food security. It is on its way to achieving sustainable mobility and lighting. However, this is not a silver bullet. If used irresponsibly, it can easily lead to chaos in the communities where it is implemented.

Food security is being threatened by the consequences of human activity resulting in scarcity of water or climate change. The latter represents the biggest challenge for agricultural systems. Transgenic crops resistant to hostile weather, infections, and with increased yields or containing additional vitamins represent a solid strategy for mitigation and adaptation.

Keywords: Transgenics, genetically modified organisms, biotechnology, food security, climate change adaptation, sustainability, biodiversity, ethical concerns, agricultural innovation, regulatory frameworks.

Author

Ivan Ransom, Protect Our Planet Movement's Sustainability Ambassador and Youth Mentor

Introduction

Historically, humans have domesticated other species to secure the survival of its own. According to MacHugh et al., it all began in the Upper Palaeolithic with the domestication of the wolf (*Canis lupus*) 15,000 years before present (YBP) in a mutualistic relationship in which they would protect and assist humans to hunt and in return our species would share food and eventually shelter them (1). Records show that in the Neolithic Period, 13,000 YBP, wheat (*Triticum aestivum*) and barley (*Hordeum vulgare*) were domesticated in societies that left behind the hunter-gatherer dynamics and settled, which allowed the domestication of cattle, 10,000 YBP, starting with sheep (*Ovis aries*), goats (*Capra hircus*), pigs (*Sus scrofa*) and humpless cattle (*Bos taurus*) (2)(1).

This accelerated evolutionary process driven by humans selecting desirable traits for their own interest was named Artificial Selection, as opposed to Natural Selection, the evolutionary process proposed by Charles Darwin in which evolution of the species is driven by the pressures exerted by their environment (3). Domestication and artificial selection shaped our relationship with other species and became the foundation of human civilization. It provided sustenance for millennia. However, in recent times, and as a result of our growing population, even with modern industrialized farming processes food security has become an issue. For this, gene editing technologies arose as a potential solution.

Gene editing techniques are revolutionizing the world as we know it, enabling the production of organisms with improved or new traits. By the development of these technologies, we have been able to transform health, food, and even urban systems. However, Genetically Modified Organisms (GMOs) are not a silver bullet. Their implementation must be carefully assessed before introducing them to the market especially those in the food industry, as will be explored in this chapter. But first, let's cover the basic concepts.

A GMO is "*an organism whose genome has been engineered in the laboratory in order to favor the expression of desired physiological traits or the generation of desired biological products*" (4). A transgenic organism, on the other hand, is an organism that has had one or more genes from a different species inserted into its genome. Thus, while all transgenic organisms are GMOs, not all GMOs are transgenic.

How is a transgenic organism created? This process encompasses multiple steps: gene isolation, vector construction, transformation, and selection. First, a gene of interest is identified and isolated from a donor organism. This gene is then inserted into a vector—often a bacterial plasmid or viral carrier—that facilitates its transfer into the recipient organism's genome. Through transformation techniques such as agrobacterium-mediated transfer, electroporation, or gene gun delivery, the transgene is integrated into the host's DNA (5)(6)(7). The modified cells are then selected and cultivated to ensure a stable expression of the desired trait.

In 1973, Herbert Boyer and Stanley Cohen created the first transgenic organism, an *Escherichia coli* (8), which paved the way for numerous applications in agriculture, medicine, and industry. This breakthrough led to the development of genetically modified crops, bioengineered insulin and recombinant proteins, demonstrating the transformative potential of genetic engineering. However, the process does not end when the transgenic organism is created.

In order to release a transgenic organism or its products into the environment or the marketplace, it has to pass over regulations which may be strict or negligible varying from country to country. Another aspect that may vary with respect to their regulations are their priorities, as some focused on the method or process by which it was created, while others focus on the potential risks resulting from the novelty of a trait entering the human or natural system (9). Canada is a good example of a coherent regulatory system. This country has a strict regulatory system which prioritizes the precautionary principle which allows them to adjust efficiently to new developments in science. Moreover, their approach fosters innovation while maintaining science-based regulatory expertise (9).

There is an ongoing debate relating to transgenic organisms, whether they are important tools to overcome present and future challenges or are they a source of potential environmental and health risks that go beyond our current understanding? In this chapter we delve into the benefits and risks of transgenics and their role in climate change and health.

The benefits of transgenics: Advancing science and society

Transgenic technology has transformed medicine, agriculture, and research, offering revolutionary solutions to global challenges in just a few decades. In the early 1920's, insulin was purified from porcine and bovine pancreases (10). Despite being an important advancement in medicine, it presented many drawbacks such as immune reactions, limited supply and purity issues (11). However, this changed in 1978, when David Goeddel and his team developed the first synthetic human insulin which was approved by the FDA in 1982 (12)(13). Nowadays, human insulin is mostly obtained from transgenic *E. coli* overcoming the issues previously stated and allowing

800 million adults worldwide to successfully treat this hormone production deficiency (14)(15).

In past centuries, there has been an epidemic of blindness and weakened immune systems of children in South and South-East Asia due to Vitamin A deficiency as food containing this nutrient was either scarce or beyond the budget for millions of people (16). To tackle this, in the year 2000, Ingo Potrykus and Peter Beyer developed a variety of rice (*Oryza sativa*), the most culturally rooted and extensively consumed crop in the region, that could provide Vitamin A to this population. And that is how the Golden Rice came to be (17). It was the result of inserting genes from daffodil (*Narcissus pseudonarcissus*) and bacterium (*Erwinia uredovora*) for beta-carotene production and to improve biosynthesis respectively (17). Another example of the potential in biotechnological innovation was the development of frost-resistant tomatoes, which were created by introducing antifreeze protein genes from the abyssal fish winter flounder (*Pseudopleuronectes americanus*) (18). However, they were never commercially released due to concerns over consumer acceptance, regulation, and effectiveness (19).

Since the early 1990s, genetic engineering efforts to enhance lipid production in microalgae have been ongoing. In 1993, Christoph Benning, then at the Institute for Gene Biology Research in Germany, began his significant work on lipid metabolism (20). His research has been crucial in understanding and manipulating lipid biosynthesis pathways to increase oil content in microalgae. Later, at Michigan State University, Benning's team developed genetic tools and conducted studies on species like *Nannochloropsis oceanica*, leading to improved lipid accumulation with the hope of producing a sustainable biofuel source (21).

What if instead of generating CO_2 footprint to build streetlights, we grow one which fixes carbon dioxide, produces oxygen, cool down the microclimate and serve as home for many species while beautifying the streets? One of the blessings of biotechnology is that the limit lies even beyond our imagination. Daan Roosegaarde, a visionary, has been exploring the path to create bioluminescent trees since the early 2010s (22).

On top of these innovative biotechnological applications comes CRISPR, which is already propelling gene editing to the next level. CRISPR (Clustered Regularly Interspaced Short Palindromic Repeats) is a revolutionary gene-editing technology that allows precise modifications to DNA. It originated from a natural defense mechanism in bacteria which use CRISPR-associated (Cas) proteins to recognize and cut viral DNA. The modern CRISPR-Cas9 system was developed as a gene-editing tool by Emmanuelle Charpentier and Jennifer Doudna in 2012 enabling targeted genetic modifications with unprecedented accuracy (23). Their discovery, published in *Science*, laid the foundation for advancements in medicine, agriculture, and biotechnology, earning them the Nobel Prize in Chemistry in 2020.

However, as previously mentioned, the implementation of GMO's is still far from perfect. While some studies support the implementation of Genetically Modified Crops (GMC) as critical tools to maximize production while minimizing environmental impact, its development has been controversial due to concerns about potential negative environmental and health effects, the corporatization of agriculture, and ethical objections to manipulating life in the laboratory (24)(25).

The risks of transgenics: Ethical and environmental challenges

Genetically Modified Crops were first implemented 31 years ago, and while we have seen the agricultural industry strengthen, it has also had its social and environmental costs (26). Back in 2013, Bennett et al. reported that GMC had grown on more than 170 million hectares across developing (52%) and developed (48%) countries, which generated substantial history and extensive data relating the economic and environmental impacts of GM crops (25). Their study showed that GMC's adoption led to more benefits than negative impacts raising a call to develop "*a less polarized debate on genetic modification, focusing instead* on *how GM technologies can complement a diversified farming system framework*". While this may hold true, the most sensible approach to transgenics should be to assess case by case rather than universally condemning or embracing them. To back this premise, we will cover the case study of Monsanto's transgenic maize and its implementation in Mexican agricultural lands.

In 1997, Monsanto released a Bt maize, engineered to resist the European corn borer using *Bacillus thuringiensis* genes and one year later the Roundup Ready maize, engineered for resistance to the glyphosate herbicide, that was part of a technology package which included the said seeds, herbicides (roundup with glyphosate) and fertilizers (27)(28). Its outstanding performance in increasing crops yields spoke for itself, making it a great success and leading to its widespread adoption particularly in the United States, Argentina, Brazil, and South Africa. However, this technology was developed for profit only, neglecting social and environmental wellbeing.

This maize contained the so called "Terminator technology", formally known as Genetic Use Restriction Technology (GURT), a genetic modification which renders all its own seeds sterile translates into farmers not being able to save some seeds to plant for the next season but generating a dependency on buying their transgenic seeds to keep producing. Furthermore, and probably the most alarming aspect, is that non-GM maize pollinated by Monsanto's Bt maize would turn sterile as well, posing a huge threat for them. As this plant is pollinated by wind, efforts to prevent this pollination are futile.

If this technology was not concerning enough, the idea of implementing it in Mexican agricultural lands is catastrophic considering that Mexico is the center of origin of maize (*Zea mays*) (29), and it can potentially drive to extinction the vast varieties of endemic maize, whose process of domestication and subsequent evolution began 10,000 years ago (30).

Moreover, the lobby of Monsanto (now Bayern) to introduce their maize in Mexico has been relentlessly pushing through international trade agreements (such as USMCA) despite Mexican government's repeatedly declining this following the precautionary principle (31). While transgenic maize is currently banned from being planted in the country, recently the pressure exerted through the USMCA led Mexico to have to import transgenic maize (or products containing it) for human consumption (32). If this battle is lost the future will not look promising for native maize in Mexico and thus, its rich germplasm, patrimony of humanity, will be at risk of being lost forever.

As biodiversity declines globally, the economic and ecological consequences of this loss are becoming more apparent with economists highlighting the long-term benefits of biodiversity protection (33). Transgenic crops, while offering potential benefits such as increased yields and pest resistance, may also contribute to genetic homogenization, disrupting natural ecosystems and threatening native species. This is especially concerning for developing nations, where agriculture remains a primary driver of economic growth and poverty alleviation (34).

Climate change and food security

Climate change is one of the greatest ecological and social challenges of our century (35)(36). Despite that, there are still debates on whether it is possible to balance sustainability with economic growth or ecological diversity with global food production as if the business-as-usual approach was still negotiable (36)(37). It has been extensively documented that human-caused climate change will affect both the availability and nutritional value of food and water as well as our capacity to distribute them fairly (37)(38).

While humans have historically adapted to changing climatic conditions, comprehensive evidence of large-scale adaptation is still limited. Effective adaptation is not entirely autonomous; it depends on informed decision-making, strategic planning, coordination, and foresight. However, various challenges such as gaps in knowledge, behavioral resistance, and market inefficiencies impede successful adaptation, highlighting the need for targeted policy interventions (39). Despite the great progress that has been achieved in tackling world hunger during the past decades, the food and water systems will keep facing an unsustainable rising demand that will make it increasingly harder to meet the needs of everyone if production and consumption patterns are not modified (38)(40).

Vermeulen et al. reported that food systems accounted for approximately 19 percent to 29 percent of global anthropogenic greenhouse gas (GHG) emissions, releasing between 9,800 and 16,900 megatons of carbon dioxide equivalent ($MtCO_2e$) back in 2008 (41). Climate change will impact agricultural productivity, farm incomes, food prices, supply chain reliability, food quality, and, importantly, food safety. Low-income food producers and consumers will be disproportionately affected due to their limited capacity to invest in adaptive technologies and institutions amid growing climate risks. While there are potential synergies between food security, adaptation, and mitigation, implementing effective solutions such as agricultural intensification or waste reduction will require careful planning to ensure equitable distribution of costs and benefits (41).

If implemented correctly, transgenics can play a key role

in ensuring global food security in the context of climate change. However, adequate implementation followed by correct communication strategies will be key to overcome people's prejudices regarding GMOs (42).

The rise in the Earth's temperature increases plant's stressors or their vulnerability to them, limiting their productivity and putting food security at risk (43)(44). Husaini & Sohail propose that cultivating genetically modified (GM) crops without chemical inputs can enable agricultural intensification while minimizing negative health and environmental effects (45). Furthermore, they sustain enhancing these crops with high-value pleiotropic genes, alongside advanced technologies such as robotics and artificial intelligence (AI), can further boost productivity. This approach to "organic-GM" agriculture would provide consumers with nutritious, agrochemical-free GM produce. These innovative practices have the potential to initiate a new era of chemical-free GM farming within Agriculture 4.0, supporting efforts to achieve the United Nations Sustainable Development Goals (SDGs) (46).

Concluding comments

Our species survival depends on the survival of other species, and it is paramount to pursue a sensible dynamic that benefits us all and, ultimately, the planet as a whole. Transgenics have allowed us to overcome many challenges regarding human health and food security and are on their way to solving others related to mobility and urban infrastructure. However, their implementation should always prioritize the best interest of societies and nature. Transgenics should not be universally embraced or rejected. A proper assessment for each case is needed to reach a sustainable future for all.

References

1. MacHugh DE, Larson G & Orlando L. Taming the past: Ancient DNA and the study of animal domestication. Annual Review of Animal Biosciences. Annual Reviews Inc.; 2017; 329–51.

2. Purugganan MD & Fuller DQ. The nature of selection during plant domestication. Nature. 843–8 (457). 2009.

3. Darwin, C. On the origin of species by means of natural selection, or the preservation of favoured races in the struggle for life. John Murray. 1859.

4. Britannica. Genetically modified organism. Britannica. 2025. https://www.britannica.com/science/genetically-modified-organism

5. Gelvin SB. Agrobacterium -mediated plant transformation: the biology behind the "Gene-Jockeying" tool. Microbiology and Molecular Biology Reviews. 2003 Mar;67(1):16–37.

6. Tsong TY. Electroporation of cell membranes. Biophysical Journal. 1991; 297–306 (60).

7. Zhang D, Das DB & Rielly CD. Potential of microneedle-assisted micro-particle delivery by gene guns: A review.

Drug Delivery. Informa Healthcare. 2014; 571−87 (21).

8. Cohen SN, Chang ACY, Boyert HW & Hellingt RB. Construction of biologically functional bacterial plasmids in vitro (R factor/restriction enzyme/trans formation/endonuclease/antibiotic resistance). PNAS. 1973. https://www.pnas.org

9. Ellens KW, Levac D, Pearson C, Savoie A, Strand N, Louter J, et al. Canadian regulatory aspects of gene editing technologies. Transgenic Research. 2019 Aug 1;28:165−8

10. Pickup J. Human insulin. British Medical Journal. 1986.

11. Bolli GB, Cheng AYY & Owens DR. Insulin: evolution of insulin formulations and their application in clinical practice over 100 years. Vol. 59, Acta Diabetologica. Springer-Verlag Italia s.r.l.; 2022; 1129−44.

12. Quianzon CC & Cheikh I. History of insulin. Journal of Community Hospital Internal Medicine Perspectives. 2012 Jan 16;2(2):18701. https://www.tandfonline.com/doi/full/10.3402/jchimp.v2i2.18701

13. Food & Drug Administration. 100 years of insulin. FDA. 2025 Apr 26 https://www.fda.gov/about-fda/fda-history-exhibits/100-years-insulin

14. Riggs AD. Making, cloning, and the expression of human insulin genes in bacteria: The path to humulin. Endocr Rev. 2021 May 25;42(3):374-380. doi: 10.1210/endrev/bn

aa029. PMID: 33340315; PMCID: PMC8152450.

15. Reuters. More than 800 million adults have diabetes globally, many untreated, study suggests. Reuters. 2024 Nov 14. https://www.reuters.com/business/healthcare-pharmaceuticals/more-than-800-million-adults-have-diabetes-globally-many-untreated-study-2024-11-13/

16. West, K. P., & Darnton-Hill, I. Vitamin A deficiency. Nutrition and Health in Developing Countries. 2008; 377–433.

17. Ye X & Beyer P. Engineering the provitamin A (β-carotene) biosynthetic pathway into (carotenoid-free) rice endosperm. Science. 2000;287(5451):303–5.

18. Eskandari A, Leow TC, Rahman MBA & Oslan SN. Antifreeze proteins and their practical utilization in industry, medicine, and agriculture. Biomolecules. MDPI AG; 2020; 1–18.

19. Lemaux PG. Genetically engineered plants and foods: A scientist's analysis of the issues (part I). Annual Review of Plant Biology. 2008; 771–812 (59).

20. Arondel V, Benning C & Somerville CR. Isolation and functional expression in Escherichia coli of a gene encoding phosphatidylethanolamine methyltransferase (EC 2.1.1.17) from Rhodobacter sphaeroides. Journal of Biological Chemistry. 1993 Jul 25;268(21):16002–8.

21. Vieler A, Wu G, Tsai CH, Bullard B, Cornish AJ, Harvey C, et al. Genome, functional gene annotation, and nuclear transformation of the heterokont oleaginous alga nannochloropsis oceanica CCMP1779. PLoS Genetics. 2012 Nov;8(11).

22. Salvatore, C. Is green illumination on the horizon? Fisher Science Education. 2014. https://www.fishersci.co m/us/en/education-products/publications/headline-discoveries/2014/issue-04/green-illumination-on-horizon.html

23. Jinek M, Chylinski K, Fonfara I, Hauer M, Doudna JA & Charpentier E. A programmable dual-RNA-guided DNA endonuclease in adaptive bacterial immunity. Science. 2012. https://www.science.org/doi/epdf/10.1126/scienc e.1225829

24. O'Brien M & Mullins E. Relevance of genetically modified crops in light of future environmental and legislative challenges to the agri-environment. Annals of Applied Biology. 2009; 323–40.

25. Bennett AB, Chi-Ham C, Barrows G, Sexton S & Zilberman D. Agricultural biotechnology: Economics, environment, ethics, and the future. Annual Review of Environment and Resources. 2013 Oct; 38:249–79.

26. Connie C. GMO food fears and the first test tube tomato. Retro Report.2013. https://web.archive.org/web/202203

06145519/https://www.retroreport.org/transcript/test-tube-tomato/

27. Eckhoff, S. R., Paulsen, M. R., & Yang, S. C. Maize. Encyclopedia of Food Sciences and Nutrition. 2003; 3647–3653.

28. WayBack Machine. Monsanto: Company history. WayBack Machine 2004-2008. https://web.archive.o rg/web/20080423174556/http://www.monsanto.com/ who_we_are/history.asp

29. Santillán-Fernández A, Salinas-Moreno Y, Valdez-Lazalde JR, Bautista-Ortega J & Pereira-Lorenzo S. Spatial delimitation of genetic diversity of native maize and its relationship with ethnic groups in Mexico. Agronomy. 2021 Apr 1;11(4).

30. Britannica. Corn. Britannica. 2025. https://www.britan nica.com/plant/corn-plant

31. Mexico News Daily. After a 4-year legal battle, Monsanto drops lawsuit against Mexico's GM corn ban. Mexico News Daily. 2024. https://mexiconewsdaily.com/news/ monsanto-drops-lawsuit-mexico-gm-corn-ban/

32. El País. México pierde el panel sobre el maíz transgénico frente a EE UU y Canadá. El País. 2024. https://elpais.c om/mexico/economia/2024-12-20/mexico-pierde-el-panel-sobre-el-maiz-transgenico-frente-a-ee-uu-y-canada.html

33. Hanley N & Perrings C. The economic value of biodiversity. Doi Foundation. 2019. https://doi.org/10.1146/annurev-resource-100518-

34. de Janvry A & Sadoulet E. Agriculture for development: Analytics and action. Annual Reviews. 2022 Jun 8.

35. Dietz T, Shwom RL & Whitley CT. Annual review of sociology. Annual Review. 2024; 9:30. www.annualreviews.org.

36. Jayachandran S. How economic development influences the environment. Annual Review of Resource Economics. 2024;10(10):25. https://doi.org/10.1146/annurev-economics-

37. Robinson GM. Globalization of agriculture. Annual Review of Resource Economics. 2018 May 31; https://doi.org/10.1146/annurev-resource-

38. Myers SS, Smith MR, Guth S, Golden CD, Vaitla B, Mueller ND, et al. Climate change and global food systems: Potential impacts on food security and undernutrition. Annu Rev Public Health. 2024; 38:53. https://doi.org/10.1146/annurev-publhealth-

39. Fankhauser S. Adaptation to climate change. Annual Review of Resource Economics. 2024 Apr 3;9:28. https://doi.org/10.1146/annurev-resource-

40. Nature. Genetics for a warming world: Nature Genetics. Nature Publishing Group. 2019; 1195.

41. Vermeulen SJ, Campbell BM & Ingram JSI. Climate change and food systems. Annual Review of Environment and Resources. 2012; 195–222.

42. Lu H, McComas KA & Besley JC. Messages promoting genetic modification of crops in the context of climate change: Evidence for psychological reactance. Appetite. 2017 Jan 1;108:104–16.

43. Kim HS, Wang W, Kang L, Kim SE, Lee CJ, Park SC, et al. Metabolic engineering of low-molecular-weight antioxidants in sweetpotato. Plant Biotechnology Reports. Springer; 2020; 193–205.

44. Ayala FM, Eurídice Hernández-Sánchez I, Chodasiewicz M, Wulff BBH & Svačina R. Engineering a One Health super wheat. Annual Review of Phytopathology. 2024. https://doi.org/10.1146/annurev-phyto-121423-

45. Husaini AM & Sohail M. Agrochemical-free genetically modified and genome-edited crops: Towards achieving the United Nations sustainable development goals and a "greener" green revolution. Journal of Biotechnology. Elsevier B.V.; 2024; 68–77.

46. United Nations. United Nations sustainable development goals. The 17 goals. United Nations. 2025. https://sdgs.un.org/goals

4

From Exposure to Policy: Tracing Human Health Risks of Marine Microplastics and their Regulatory Landscape

Abstract

Marine plastic pollution, particularly microplastics (MP), is a growing global crisis that threatens ocean ecosystems, biodiversity, and human health. Microplastics persist in marine environments and contaminate oceans, freshwater systems, and the food chain. Recent studies show microplastic accumulation in human tissues, raising concerns about links to chronic diseases such as cancer and cardiovascular conditions. However, their long-term health impacts remain poorly understood. Key contributors to the crisis include inadequate waste management, limited public awareness, and improper disposal practices. Climate change intensifies microplastic pollution by altering ocean currents and accelerating plastic degradation. Rising sea surface temperatures, cyclones, and coastal flooding contribute to plastic redistribution, while ocean acidification and coral reef degradation weaken natural filtration, increasing MP retention in food chains and heightening exposure risks.

This chapter provides a comprehensive overview of MP pollution, analyzing its sources and pathways to develop effective solutions. It reviews advancement in detection technologies and emphasizes the need for a standardized methodology to improve consistency and accurately identify and quantify microplastic contamination in food and water. The chapter also highlights the importance of improving health impact assessments and strengthening marine monitoring programs to ensure accurate data collection. Additionally, it advocates for robust regulatory frameworks and better waste management to reduce microplastics spread, calling for a coordinated global response to tackle this pressing environmental challenge.

Keywords: Marine pollution, plastics, microplastics, chronic diseases, climate change, chronic diseases, awareness.

Authors

Chime Youdon, Research Fellow, Head, Resilience and Sustainability of Ocean Resources (RSOR) Cluster, National Maritime Foundation, India

Soham Agarwal, Associate Fellow, Resilience and Sustainability of Ocean Resources (RSOR) Cluster, National Maritime Foundation, India

Introduction

Fifty to seventy-five trillion pieces of plastic and microplastic populate the world's oceans, a number which grows by the day (1). Albeit developed in 1907, the large-scale production of synthetic plastics during the 1940s and 1950s led to their proliferation in human society and consequently our environment due to their durability, cost-effectiveness, and manufacturing versatility (2). As a result, global plastic production skyrocketed from just two million metric tons in 1950 to over 400 million metric tons by 2015. By 2021, cumulative production exceeded a staggering nine billion metric tons. If current trends persist, this figure is projected to reach an alarming 33 billion metric tons by 2050 and it is estimated that by this time there will be more plastic than fish in the sea (3). This dramatic boom in plastic production has come at a profound environmental cost.Inadequate waste management systems have resulted in nearly 60 percent of all plastic waste being either disposed in landfills or directly released into the environment (4). Due to their synthetic nature, these polymers persist for centuries, with vast quantities ultimately making their way into marine ecosystems.

Since the 1980s, plastic waste has become the dominant form of marine litter (5). Each year, between 12.8 to 14.8 million metric tons of plastic enter the marine environment, much of it accumulating in large gyre-driven garbage patches such as the Great Pacific Garbage Patch, which spans roughly 1.6 million km^2 (6)(7). A significant portion of this plastic enters the ocean through rivers and breaks down into microplastics ($1\mu m-5$ mm) and even smaller nanoplastics as it degrades

(8). Coastal regions across the world have emerged as plastic pollution hotspots. For example, in India, coastal states such as Gujarat (Gulf of Kachchh and Gulf of Khambhat) Mumbai, Kerala, Goa, Kanyakumari, Andaman and Nicobar Islands along with fishing harbors in Kerala and Karnataka – emerge as hotspots, fueled by industrial discharge, tourism, fishing, and oceanic currents (9).

Microplastics were first identified as a major ecological issue in 2004 (10). Since then, they have become a focus of environmental research due to their pervasive presence and associated health risks. These microscopic particles not only infiltrate ecosystems but also act as carriers for toxic pollutants and invasive species, exacerbating environmental damage (11). Between 2015 and 2022, concentrations of plastic in the ocean rose sharply—from 2.9 kg to 14.2 kg per square kilometer— while microplastic hotspots expanded tenfold (12). Today, microplastics are estimated to outnumber the stars in our galaxy by a factor of 500 times, posing an unprecedented threat to biodiversity, ecosystem resilience, and human health (13). Economically, microplastic pollution causes $8 billion annual damage to marine industries, while healthcare costs may rise due to links with chronic diseases (14).

Addressal mechanisms includes policy and regulatory measures at the global, regional, and national level— such as the developing United Nations Environment Assembly's (UNEA) Global Plastics Treaty and Europe's Restriction of Microplastics Initiative under the 2018 Plastics Strategy— are evolving to curb this crisis (15). However, the lack of standardized detection methodologies hampers impact assessments (16).

It begins by identifying microplastic pathways, sources, and transformation in the ocean and assessing the current state of research of their impact upon human health. Thereafter, policy and regulatory frameworks at the global, regional, and national level on microplastics are mapped. This chapter seeks to highlight the importance of improving health impact assessments and strengthening marine monitoring programs to ensure accurate data collection. The case is made for a robust, coordinated national implementation strategy supported by a legal framework for the reduction and spread of microplastics.

Microplastic pathways, sources, and transformation in the ocean

Categories of microplastics based on origin: Primary versus secondary

Microplastics (<5 mm) are classified as primary or secondary based on their origin. Primary microplastics, i.e., those that are intentionally manufactured for products like cosmetics, textiles, and industrial abrasives contribute 15–31 percent of oceanic microplastic loads, with key sources including synthetic textile laundering (35%) and tire abrasion (28%) (17). Secondary microplastics, comprising 69–85 percent of oceanic microplastic load, result from the degradation of larger plastics (e.g., single-use plastics, fishing gear) through photodegradation (ultraviolet exposure), mechanical degradation (wave action), and limited biological degradation (microbial colonization) (18). This classification oversimplifies their interconnected lifecycle as primary microplastics can degrade further into smaller secondary forms and data variability across studies undermines precision. This diverts the focus from systemic drivers like industrial production and consumer behavior which require integrated mitigation strategies.

Mode of entry: Land-based versus ocean-based

Microplastics enter the ocean predominantly from land-based sources, accounting for 80–90 percent of marine plastic pollution (19). Among land-based contributors, synthetic textiles are a major contributor. According to UNEP (2018), microplastics released from textiles washing account for around nine percent of total environmental discharges. However, when accounting for the fact that roughly half of the microplastics from tire wear, road markings, and urban dust are retained in soils or along riverbanks, the share of textile-derived microplastics reaching the ocean increases to 16 percent (20). Ocean-based sources such as discarded fishing gear, waste from ships, offshore platforms debris, and aquaculture operations contribute the remaining 10–20 percent of marine plastic pollution (21). However, research into marine microplastics is evolving with newer technologies and methodologies constantly updating our understanding on the sources of marine microplastics. Further, these figures differ regionally depending upon the level and type of discharge from the littorals of seas and oceans.

Figure 1

Mode of entry of marine plastic

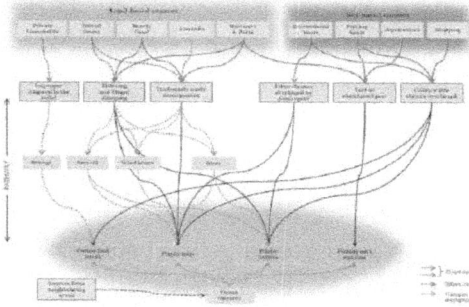

Source: Joana Mira Veiga et al. Identifying sources of marine litter – Technical report. European Commission. 2016.

Microplastics reach marine environments through multiple pathways—waterborne (e.g., rivers), airborne (e.g., wind deposition), and direct maritime discharge (Figure 2). Rivers and surface runoff act as key transport pathways with an estimated 90 percent of ocean plastics originating from just 10 rivers – nine of which are in Asia (22). Once in the ocean, MPs undergo physical, chemical, and biological transformation, fragmenting into smaller particles, releasing toxic additives, and becoming increasingly bioavailable (23). Climate change exacerbates these processes. Altered ocean currents redistribute MPs to previously uncontaminated areas while rising sea temperatures accelerate degradation and the leaching of toxic compounds intensifying interactions with marine organisms (24). Additionally, coastal flooding flushes land-based MPs into aquatic systems increasing their persistence in food webs (25).

Figure 2
Key sources and pathways of microplastics in the Ocean

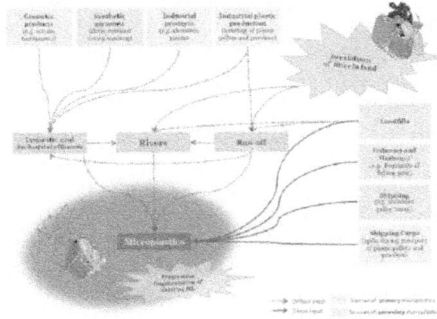

Source: Joana Mira Veiga et al. Identifying sources of marine litter – Technical report. European Commission. 2016.

The long-term impact of microplastics on agriculture, fisheries, and livelihoods remain unclear, but concerns persist over their toxicity and ecological risks. As microplastics degrade, they can leach harmful chemicals and carry pathogens, potentially triggering inflammation, toxicity, and broader environmental harm (26). Climate-induced fragmentation and dispersion further amplify the transport and bioaccumulation of POPs, heightening toxicity within marine food webs (27). However, this linear source-to-impact model often overlooks regional variability and socioeconomic drivers such as poor recycling infrastructure. Addressing these intertwined challenges calls for interdisciplinary frameworks that integrate ecological, social, and economic dimensions. The complex interplay of MPs, POPs, and climate change underscores the urgent need for adaptive science-based strategies to mitigate their combined threat to marine ecosystems.

Pathways of microplastic exposure in humans

Humans are exposed to microplastics through multiple pathways primarily via ingestion, inhalation, and dermal contact.

Figure 3
Potential pathways of marine microplastic into our seafood

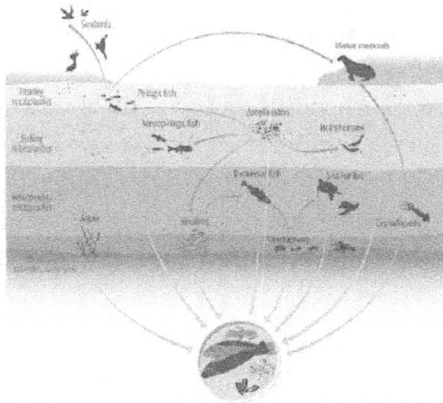

Source: UNEP 2021. Drowning in plastic- marine litter and plastic waste vital graphics. GRID-Arenda.

The most significant exposure routes include:

Ingestion through seafood

Microplastics are ingested by all kinds of living organisms ranging from plankton, fish, and large mammals in the marine environment. Since the 1960s, it is documented that microplastics are ingested by marine species that affects seabirds, mammals, turtles, and fish, permeating food webs via direct ingestion and trophic transfer (Figure 3) (28). Filter-feeders like mussels and oysters are key contributors to human dietary exposure to microplastics which carry along with them pollutants like persistent organic pollutants (POPs) and heavy metals (29)(30). Beyond seafood, microplastics contaminate salt, honey, drinking water, dairy, and beverages making exposure unavoidable (Table 1). Inhalation and dermal contact further increase exposure, particularly in urban areas.

Health risks associated with microplastic exposure

The small size and chemical composition of microplastics facilitate their bioavailability leading to potential physiological and toxicological effects. Exposure to microplastics has been associated with cellular and tissue damage, chemical toxicity, endocrine disruption, immune dysfunction, and potential carcinogenic effects.

Cellular and tissue damage: Microplastics and nanoplastics penetrate membranes accumulating in organs (e.g., liver, brain) and cause oxidative stress, inflammation, and mitochondrial dysfunction (31).

Toxicity and chemical exposure: Microplastics adsorb and transport toxic substances such as polychlorinated biphenyls (PCBs), bisphenol A (BPA), phthalates, and heavy metals leading to endocrine disruption, metabolic disorders, and carcinogenesis (32). Upon ingestion, microplastics may leach these chemicals into the digestive system contributing to systemic toxicity and inflammation (33). The bioaccumulation of these toxicants in organs raises concerns about their long-term health impacts including neurological and immune system disorders which are associated with endocrine disruption, metabolic disorders, and inflammation (34). Some microplastics are small enough to cross the intestinal barrier and enter the bloodstream potentially accumulating in organs and inducing oxidative stress (35).

Endocrine disruption and reproductive effects: Microplastics and their associated chemicals interfere with hormonal regulation and reproductive health. Studies indicate that exposure to plastic-derived pollutants can reduce sperm quality, impact embryonic development, and alter gene expression related to fertility. Additionally, plastic additives such as BPA have been linked to developmental disorders and transgenerational health effects raising concerns about long-term reproductive consequences (36).

Potential carcinogenic effects: Chronic exposure to microplastic-associated carcinogens (e.g. PAHs) may increase cancer risk (37).

Immune system dysfunction: Microplastic ingestion and inhalation have been linked to altered immune responses potentially leading to immune suppression or hyperactivation. Experimental studies on marine life indicate that microplastic exposure induces pro-inflammatory cytokine production and disrupts immune cell function (38). These findings suggest that chronic exposure could weaken immune resilience, making individuals more susceptible to infections and inflammatory diseases.

Neurological and developmental concerns: Exposure to

microplastic-associated neurotoxicants including BPA and phthalates has been associated with cognitive impairment, behavioral changes, and neurodevelopmental disorders (39). There is growing concern that maternal exposure to microplastics during pregnancy could impact fetal brain development, potentially leading to neurodevelopmental deficits. As microplastics continue to accumulate in the environment and the food chain, the potential risks to neurological health warrant further investigation.

Empirical evidence of microplastic contamination in human food sources

Several studies have documented the presence of microplastics in various dietary sources, reinforcing concerns about human exposure (Table 1). As microplastics continue to accumulate across environmental compartments and infiltrate the food chain, growing concerns have emerged regarding their potential impacts on neurological health. Early studies suggest that microplastics may induce neurotoxic effects through mechanisms such as oxidative stress, inflammation, and disruption of the blood-brain barrier. However, evidence remains limited and fragmented, underscoring the urgent need for systematic, multidisciplinary research to understand the extent and nature of these risks. In this context, improving the quality and scope of health impact assessments becomes critically important—not only to clarify potential neurological and systemic effects, but also to inform evidence-based regulatory responses. Equally vital is the strengthening of marine monitoring programs that can consistently track microplastic presence and behavior in aquatic systems. Together, robust health assessments and enhanced monitoring will help generate reliable, high-resolution data that is essential for guiding targeted policy interventions and safeguarding public health in the face of this emerging environmental threat. Key findings include:

Table 1		Summary of species documented to have ingested plastics			
Category	Study	Findings	Methodology	Health impact	Country
Fish	Comprehensive seafood study (40)	MPs in fish: 0.96 ± 0.08 MP/fish (herring, sardine, whiting, flathead)	Microplastic characterization	Seafood as MP exposure pathway	Australia
	Golden anchovy study (41)	MPs in anchovies: 6.78 ± 2.73 MP/fish (fibers, films, pellets)	Tissue microplastic identification	Risk from whole-fish consumption	India
Canned Fish	Karami et al. (2018) (42)	Micro- and mesoplastics in canned sardines and sprats	Microscopic analysis	Increased exposure from canned seafood	Malaysia
	Canned fish study (43)	MPs in canned mackerel and tuna: 1.28 ± 0.04 MP/g (PET, PS, PVC)	Canned fish analysis	Higher exposure from processed seafood	Iran

Shellfish	Van Cauwenberghe & Janssen (2014) (44)	Microplastics in mollusks. Avg 11,000 MPs/year (European consumption)	Soft tissue analysis	Gastrointestinal/ systemic exposure	Japan
Shrimp	Indian white shrimp study (45)	MPs in shrimp: 0.04 ± 0.07 MP/g wet weight	Spectros copy analysis	Exposure through shrimp consumption	India
Seaweed	Nori seaweed study (46)	MPs in nori seaweed: 1.8 ± 0.7 MP/g	Polymer/ size analysis.	Contamination in seaweed	China
Sea Salt	Chinese study on salt (2015) (47)	MPs in sea salt: 550–681 particles/kg (fibers, fragments)	Micropla stic content analysis	Regular MP exposure via salt	China

Source: Compiled by authors

Impact of microplastic characteristics on marine ecosystems and human food systems

Microplastics vary in size, color, and shape, influencing their environmental fate, interactions with marine biota, and implications for human food systems.

Size: Bioavailability and trophic transfer

Microplastic size, ranging from nanoplastics (<1 μm) to 5 mm, governs transport, bioavailability, and toxicity. Smaller particles, with higher surface-area-to-volume ratios, are more mobile, disperse widely via ocean currents, and adsorb contaminants like polychlorinated biphenyls acting as pollutant vectors (40). Larger microplastics (1–5 mm) are ingested by filter feeders (e.g., mussels) and predators (e.g., fish), causing physical blockages and reduced feeding efficiency (41). Smaller particles (<100 μm) penetrate cellular membranes, accumulate across trophic levels, and cause transgenerational effects such as reduced zooplankton reproduction (42). In human food systems, smaller microplastics in seafood (e.g., shellfish, finfish) translocate into edible tissues, pose health risks (43).

Color: Selective ingestion and toxicity

Microplastic color (e.g., white, blue) impacts ingestion and persistence. Darker plastics degrade faster under ultraviolet light, while lighter ones persist (44). Bright colors (e.g., red) mimic prey and increase ingestion by fish, while transparent particles attract filter feeders (45). Species-specific preferences, like larval fish favoring blue — vary (46). In seafood, colored microplastics introduce toxic additives (e.g., dyes), risking chemical transfer to humans.

Shape: Ingestion, transport, and retention

Microplastic shapes—fibers, fragments, spheres, films, foams—affect hydrodynamics and biological uptake. Fibers, often from textiles, entangle in gills or digestive tracts of marine species (e.g., crustaceans), while fragments and spheres are readily ingested due to compact shapes

(47). Fibers prolong gut retention, increasing contaminant release, whereas spheres translocate efficiently across tissues, enhancing bioaccumulation (48). In human food systems, fibers dominate in shellfish and fragments in finfish reflecting habitat differences.

Challenges in microplastic detection, characterization, and regulation

The pervasive nature of microplastics and nanoplastics in environmental and biological systems poses significant scientific and regulatory challenges. Their physical and chemical heterogeneity—spanning sizes (<1 µm to 5 mm), shapes (fibers, fragments, spheres), colors, polymer types (e.g., polyethylene, polystyrene), and degradation states—complicates detection, quantification, and risk assessment. These challenges are compounded by the absence of standardized methodologies, regulatory gaps, and limitations in current analytical technologies hindering a comprehensive understanding of microplastic pollution and its implications.

Challenges in detection and characterization

The accurate detection and characterization of microplastics in environmental matrices (e.g., water, sediments, biota) are hindered by diverse particle properties and chemical additives (e.g., plasticizers, flame retardants). Existing microscopy techniques such as Optical, Scanning Electron (SEM) and Atomic Force Microscopy (AFM) offer morphological insights but are labor-intensive, error-prone, and lack chemical specificity (49). Spectroscopy methods such as Fourier Transform

Infrared (FTIR) and Raman Spectroscopy enable polymers identification but face limitations. For instance, FTIR struggles with particles below 20 µm, missing nanoplastics, while Raman is highly sensitive to fluorescence interference from organic matter (50)(51). Advanced techniques like Micro-FTIR Spectroscopy and Micro-Raman improve resolution but are time-consuming and require expertise, limiting scalability (52). Mass spectrometry-based methods (e.g., Pyrolysis-GC-MS) provide detailed chemical profiling but are destructive, costly, and unsuitable for routine monitoring (53). These methodological limitations are compounded by the dominance of non-plastic particles. For example, a study found that only 1.4 percent of 404 sediment particles analyzed via FTIR were microplastics, while 96 percent were quartz, highlighting the challenge of distinguishing plastics from natural substrates (54). The absence of standardized protocols for isolation, identification, and quantification limits comparability across studies, hindering spatial and temporal trend analysis.

Inconsistencies in sampling and quantification: The lack of uniform protocols in sampling and quantification across environments (marine versus terrestrial), depths and particle sizes results in irreproducible findings (55). Additionally, inadequate contamination control measures, such as insufficient precautions against airborne microplastic deposition during collection and analysis, further compromise data integrity (56). Natural degradation and transport variability, driven by factors such as polymer weathering rates, ocean currents, and sedimentation processes further obscure long-term pollution trends by affecting microplastic detectability (57).

These inconsistencies prevent the establishment of baseline pollution levels and accurate risk assessments, highlighting the urgent need for standardized methodologies that account for environmental and analytical variables.

Emerging technologies for improved detection: Emerging technologies are advancing microplastic detection by improving efficiency, accuracy, and scalability. Machine learning and AI-driven tools such as PlasticNet have demonstrated over 95 percent identification accuracy, significantly reducing human error and expediting large-scale data processing (58). However, their effectiveness depends on high-quality training datasets, limiting their reliability for novel or degraded microplastic particles. Portable detection tools—including fluorescence-based sensors, portable FTIR, and Raman spectroscopy—enable field-based analysis, reducing dependence on laboratory-based methods. While these technologies hold promise, validation across diverse environmental matrices is essential to ensure reliability. Additionally, scalability and cost barriers remain significant challenges, restricting widespread adoption.

Policy and regulatory gaps: Policy frameworks lag behind scientific progress due to insufficient data and inconsistent methodologies and the absence of global monitoring standards. Most policies focus on microplastics such as bans on their primary sources, while microplastic contamination in food, water, and ecosystems remains largely unregulated.Data gaps on long-term ecological and human health impact partially due to analytical challenges—delay proactive policy and regulations perpetuating a reactive approach rather than proactive approach. These gaps reflect a broader disconnect between scientific urgency and regulatory inertia. A more detailed exploration of these gaps will be undertaken here forth.

Policy and regulatory frameworks

Policy and regulatory mechanisms are an effort to respond to the challenges posed by marine litter especially microplastics. With developing literature and better scientific understanding of the environmental, ecological, and economic impacts, both voluntary and binding mechanisms are being developed at the international, regional, and national levels to address the lifecycle of plastic in attempts to address the proliferation of microplastics in the marine environment (59). The regulatory structure has evolved to distinctly address primary and secondary microplastics.

Primary microplastics

The most recent regulatory mechanism to address primary microplastics is seen within the European Union (EU). While it is acknowledged that primary microplastics represent a relatively smaller proportion of microplastic in the sea, yet they form the only type of microplastic which has been directly and explicitly regulated (60). This is because they are relatively easy to address and can allow for targeted action.

Drawing from the European Strategy for Plastics in a Circular Economy, and the Zero Pollution Action Plan — which mandates a 30 percent reduction in microplastics release into the sea by 2030 — the European Commission released Commission Regulation (EU) 2023/2055 (61)(62) . This legislation introduced "synthetic polymer microparticles" (SPM) within Annex XVII of the Regulation concerning the Registration, Evaluation, Authorization and Restriction of Chemicals (REACH), which is the primary EU law to protect human health and the environment from the risks posed by chemicals (63). Inclusion of SPMs into REACH has enabled their regulation at a European Union level through which a general ban has been imposed on the "placing on the market" of SPMs, either on their own or intentionally introduced into "mixtures" (64). Size limits are established to gauge applicability of the restriction.This includes particles equal to or less than 5 mm in diameter but also includes those particles the "length of which is equal to or less than 15 mm and their length to diameter ratio is greater than 3" when found in a concentration equal to or greater than 0.01 percent by weight of the mixture. Particles up to 15 mm in length are brought within the scope of the restriction for those which are used for reinforcement of adhesives and concrete which are "very persistent and contribute to the identified risk" (65). The minimum concentration value of 0.01 percent is identified as it is the minimum level where SPMs can confer a sought-after characteristic to the product and may be intentionally added. Therefore, these SPM (polymers as defined in the regulation) may not be "supplied or made available...to a third party" (including imports within the Union) either as themselves or within a product to which they are added to alter its properties (66).

The aim is to capture all those cases which inevitably lead to the release of these SPMs into the environment. It is estimated that this restriction will lead to a cumulative emission reduction of nearly 500,000 tonnes of microplastics over a 20-year period following the implementation of the regulation (67). While this may sound significant, it must be compared to 200,000 – 500,000 tonnes of microplastic released into the marine environment each year from the textile industry alone (68).

This can stem from the extremely limited scope of this regulation with significant derogations from this ban which in any case is sought to be implemented gradually over a period of time. In cases where the product does not release SPMs, their release can be prevented/minimized or regulated by other legislation are excluded from the application of this ban (69). There are two significant exclusions from these restrictions.

First, the scope of the term 'SPMs' is restricted to include only solid, synthetic/chemically modified, non-degradable, and insoluble polymers given the lower environmental impact of liquid, natural, degradable, and soluble polymers. This also promotes investment in developing environmentally-friendly polymers. Recent research, however, has highlighted that "biodegradable" plastics do not do so in the marine environment as they require specific conditions such as heat and microbes which otherwise are not prevalent at sea (70). Even if they do degrade, it is often at an extremely slow pace, or they degrade into smaller and more persistent particles (71). Only specific types of plastics have shown to be biodegradable by microbes in seawater, but legislation does not make any distinction between the two. This highlights the importance of informed decision-making and bridging the gap between science and regulation.

Second, is the permitting of derogations from the ban for those SPM "for use at industrial sites" among other uses and those which can be contained from release either by technical means are permanently modified at end use, or permanently incorporated in a solid matrix at end use (72). For these categories of SPMs, instructions for use and disposal (IFUD) are to be provided to industrial downstream users and end users and estimated SPM emissions are to be reported each year by the SPM supplier (73). The rationale is that such instructions can prevent or at least significantly minimize SPM release into the environment given that industries that are well-regulated will follow and implement the IFUD (74). However, this presumes that the supplier is aware of the different emission pathways for microplastics. While reporting requirements place an obligation on the supplier to undertake emission assessments, significant gaps in research and understanding may preclude accurate reporting and disposal mechanisms, making the framework inefficient. This demonstrates the challenges and competing interests often must be balanced by policymakers when developing legislation on such issues. Therefore, while this legislation may capture consumer products to which SPMs are intentionally added, its industrial use has currently been subjected to IFUD and reporting requirements. This broadly corresponds to the legislation in the United States of America attempting to restrict the manufacture and distribution of microbeads in "rinse-off cosmetics" under the Microbead-Free Waters Act 2015 broadly limited to cosmetics and non-prescription drugs (75).The objective of this Act is to prevent run-off into lakes and oceans of microplastics which cannot currently be filtered and not to address microplastic impact on human health. This stems from the fact that research on the impact of microplastics on marine ecosystems is better documented than the impact of microplastics on human health.

Hence, legislation is designed in such a way that it aims to restrict specific emission pathways and reduce impact on human health indirectly. A notable example may be the exception in the EU regulation for fiber-like particles in concrete which are incorporated in a solid matrix at end use. However, research suggests that concrete products will likely emit microplastics in the future due to erosion and demolition via leaching and airborne dust (76). While options such as the use of high strength to ultra-high strength concrete and surface hardening agents exist, they do not necessarily reflect in the legislation. Hence, research is required not only to better understand the impact on human health, which may potentially lead to more stringent regulation, but also to frame better law.

Secondary microplastics

Since secondary microplastics are formed due to the fragmentation and degradation of larger plastics, addressing this source of microplastics would requires addressing the problem of marine litter as a whole. The challenges and competing interests that were faced while regulating intentional microplastics attain an order of magnitude of significant proportion as the whole plastic industry needs to be addressed. Further, lack of viable cost-effective alternatives to plastic precludes an outright prohibition on plastic production and use. Therefore, piecemeal restrictions are being discussed and negotiated to identify the least restrictive measure with the greatest impact. Greater emphasis is placed on addressing the entire plastic lifecycle to promote plastic circularity and prevent leakage into the environment (77).

The global community has committed to adopting an international legal binding instrument on plastic pollution including in the marine environment through Resolution 5/14 of the United Nations Environment Assembly in 2022 (78). This treaty, when adopted, would be the first global legally binding instrument explicitly addressing plastic pollution. Current attempts, including through this treaty, are on developing a mix of voluntary and binding approaches at the global, regional, and national level. Since the majority of microplastics in the oceans come from land-based sources, addressing upstream activities including manufacturing, use, and waste management within states has become an important tool in the battle against microplastics. However, the principle of sovereignty of states and their right to pursue their economic development, especially for developing economies requires that these measures are non-intrusive and facilitative in nature. Any measure which places an undue burden on the states is unlikely to be agreed to and will affect the effectiveness of global measures. Since the oceans are a global common, leakages of plastic from any source will have an impact on everybody. It is in this context that voluntary mechanisms play an important role in balancing these competing interests.

Regulation of plastic and consequently microplastics does not feature as the primary objective of any international treaty. Rather, fragmented measures which have as their objective the prevention of pollution, protection of marine biodiversity and regulation of chemicals find application to the issue of plastics. Each of these treaties (summarized below) has a differing scope of application and seek to address the problem at the source.

Table 2	International legally binding mechanisms on microplastics		
Legally binding norms/mechanism			
Treaty	Provision	Theme	Remarks
United Nations Convention on the Law of the Sea 1982 (UNCLOS) (79).	Article 192	General obligation on states to protect and preserve the marine environment	
	Article 194	Measures to prevent, reduce, and control pollution of the marine environment	Applicable to all sources of 'pollution'. States are to do so in accordance with their capabilities and shall endeavor to harmonize their policies on a global or regional basis acting especially through competent international organizations.
	Article 1 (4)	Definition of pollution. *"Introduction of... substances into the marine environment which results or is likely to result in harm to marine life, hazards to human health..."*	The broad definition of pollution under UNCLOS of will likely include microplastics Therefore, while not explicitly mentioned, documented research on the deleterious impact of plastics on marine life and human health brings it within the ambit of the states' obligations to prevent or reduce their introduction into the marine environment.

	Article 207	Obligation to adopt laws and regulations to prevent, reduce, and control pollution from land-based sources including rivers.	Explicitly identifies land-based sources of pollution. Reiteration of obligation to adopt laws and harmonize policies.
	Article 211	Pollution from vessels	Establishment of global norms and effective exercise of flag state and port state jurisdiction to control pollution from vessels.
	Article 210	Pollution from dumping	Duty to regulate
	Section 6	Enforcement	A distinct obligation on states to undertake enforcement measures for the laws undertaken to prevent pollution.
International Convention for the Prevention of Pollution from Ships (MARPOL)	Annex V Regulation 3	Prohibition of plastic disposal into the sea	Adopted by the International Maritime Organization acting on the mandate on the 'competent international organization' under UNCLOS

			Addresses a major sea-based source of secondary microplastic in the oceans i.e., from ships including fishing vessels. Explicitly prohibits the disposal into sea of all plastics including synthetic fishing nets and plastic garbage bags from all ships including plastic mixed with other garbage
Convention on the Prevention of Marine Pollution by Dumping of Wastes and Other Matter 1972 (London Convention)	Article 3	Regulation of 'deliberate disposal' at sea of waste (80).	Addresses intentional dumping as opposed to the discharge into the sea of waste during the normal operations of vessels regulated under MARPOL. This treaty, however, has not received widespread ratification with currently only 87 state parties (81).
Convention on Biological Diversity (CBD).	Decision XIII/10 adopted by its Conference of Parties (COP) Paragraph 8 (f)	Addressing the impacts of marine debris on marine and coastal biodiversity (82).	While the CBD has as its primary objective the conservation of biodiversity in general, this decision focuses explicitly on marine debris impact on marine biodiversity.

			It records as a priority action for states to "*assess whether different sources of microplastics that include both primary and secondary microplastics and different products and processes, are covered by legislation, and strengthen, as appropriate, the existing legal framework so that the necessary measures are applied, including through regulatory and/or incentive measures to eliminate the production of microplastics that have adverse impacts on marine biodiversity*" (83).
Stockholm Convention 2001		Protection of human health and the environment from persistent organic pollutants (POPs)	Plastic regulation is limited to the extent of POPs produced and used in the production of certain types of plastics (e.g.: polychlorinated biphenyls (PCBs) as opposed to the production and use of plastics (84).
Basel Convention	Partnership on plastic waste was established in 2019 (85).	Transboundary movement of hazardous and 'other wastes' mandating state parties to ensure adequate disposal facilities for environmentally sound management (ESM) of hazardous and other wastes	The goal of the partnership is to improve the ESM of plastic waste in the context of addressing and reducing its transboundary movement.

Source: Compiled by authors

Voluntary mechanisms

The primary purpose of these mechanisms is to play an enabling role to allow states, through regional cooperation, to implement national and regional plans to address plastic pollution. While there is a litany of voluntary and non-profit programs to tackle marine litter, a few significant global and regional mechanisms are listed below.

Table 3	Global voluntary mechanisms on microplastics	
S. No.	Mechanism	Objective
1.	The Global Program of Action for the Protection of the Marine Environment from Land-Based Activities (GPA). - Led by a steering committee with secretarial services by the United Nations Environment Program - Seeks to implement the commitments of the 1992 United Nations Conference and Agenda 21 - Manila Declaration on Furthering the Implementation of the GPA included marine litter within the ambit of the GPA in 2012 (Manila Declaration) (79). -Works actively with the Regional Seas Program.	Established in 1995 as an intergovernmental mechanism that issues guidelines for addressing land-based sources of pollution. The GPA sought to identify specific problems for specific regions and recommend priorities for action (80). Enable development of regional action plans with steps towards harmonization of standards, capacity building, and contingency planning (81).

2.	**Global Partnership on Marine Litter (GPML)** - Established in 2012 after the Manila Declaration with a mission to protect the global environment and human wellbeing by addressing marine litter. - Voluntary open-ended partnership for state and non-state actors alike. -Framework documents were adopted in October 2018 (82). -Managed by a steering committee drawn from partners with secretarial services provided by UNEP.	Seeks to: a) create an informed global community working together; b) eliminating discharges; and c) carrying out targeted removal. Aim is to provide a platform for knowledge-sharing between all stakeholders including the private sector, NGOs, and the civil society. More specifically, GPML seeks to a) reduce the plastic in the ocean through improved design; b) application of the 3R's principle to plastics; c) promote circular production cycles; and d) maximization of resource efficiency.
3.	**Regional Seas Program (RSP)** -Established in 1974 by the UNEP -Currently it covers 18 regions, seven of which are directly administered by the UNEP (1), seven of which are established under the UNEP, but non-UNEP administered (2), while four are independently established (3). -The program is underpinned by a 'regional convention' and protocols on specific issues for each region which are collectively referred to as Regional Seas Conventions and Action Plans (RSCAPs). -Each convention and land-based protocol (if one exists to address marine litter) has differing obligations including their scope of application (4). The Mediterranean, the Western Indian Ocean, and the East Asian Sea are one of the few RSCAPs to have a dedicated protocol on Marine Litter (83).	Primary objective to address the "*accelerating degradation of the coastal and marine environment through a shared seas approach*" (84). Consequently, it seeks to build regional frameworks for cooperation, management, and protection of "shared seas". The convention sets out broad obligations for states with protocols and annexes detailing specific plans of action. However, not all protocols have entered into force and are not binding in nature with many Marine Litter Action Plans yet to be finalized (85).

4.	**Clean Seas Program[106]** -Established by the UNEP in 2017 -Mandate extended with the launch of Clean-Seas 2.0 -Flagship program is the CounterMEASURE project administered by UNEP and funded by the Government of Japan. -The project seeks to map out the places and manner in which plastic waste enters waterways, with the Ganga and Mekong rivers as pilot projects. It utilizes tools of citizen science, drone imaging, machine learning, and GIS algorithms to identify and map plastic waste (including microplastic) hotspots (87).	Primarily focused on the elimination of single-use plastic in the first edition. The second edition will focus on advocacy and promote urgency for action by using data-collection tools to identify key sources, pathways and hazards.
5.	**FAO Code of Conduct for Responsible Fisheries**	Voluntary instrument which governs issues related to fishing activities. This includes abandoned, lost, or otherwise discarded fishing gear including the development of "selective and environmentally safe fishing gear and practices".
6.	**The Cross Industry Agreement** Voluntary collaboration of five European industry associations representing the global value chain of the textile industry including manufacturing and maintenance (88).	Aims to leverage industry expertise to deploy solutions within the textile industry including in design, washing and use, and wastewater treatment.

Source: Compiled by authors

Figure 4

Summary of the legal framework applicable to microplastics

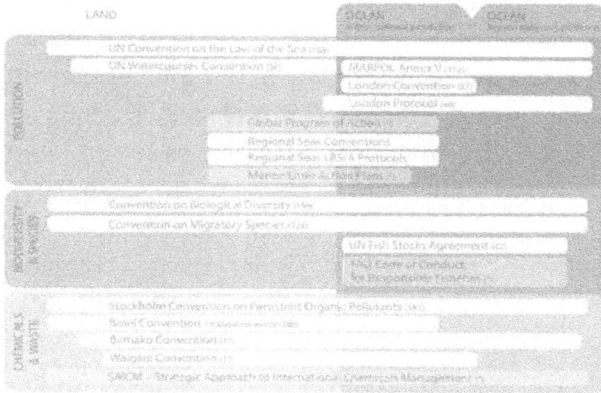

Source: United Nations Environment Program, "combating marine plastic litter and microplastics: An assessment of the effectiveness of relevant international, regional, and subregional governance strategies and approaches."

Tracing legal and policy initiatives to tackle marine litter demonstrates that microplastics do not fall in a regulatory void. While there is no specific legislation concerning plastics/secondary microplastics, the general provisions in UNCLOS are broad enough to encapsulate within their scope microplastics. However, it must be demonstrated that microplastics are indeed harmful to the marine environment and are a "land-based" source of pollution. That is why the initial focus of the international community was to establish the deleterious effect of plastic waste on the marine environment. Similarly, the impacts of microplastics on matters which were the subjects of other international conventions such as the one on protection of biological diversity and regulation of transboundary waste had to first be sufficiently established. It was only in 2012, (8 years after microplastics were identified as a major ecological concern) after it was recognized in the Manila Declaration that marine litter is an underestimated problem with profound effect on marine ecology and human health that global action was sought to be catalyzed. Notable was the recognition that land-based sources contribute a significant portion of marine litter.Therefore, action must be taken at the national level. This is reaffirmed by the fact that in 2014, the first United Nations Environment Assembly (UNEA) resolution on this subject — out of a total of four — "noted with concern" this issue was recognized as a compounding nature of the problem due to the formation of secondary microplastics (79).

However, even though the international community identifies the problem, the question of how to address it remains. This is a problem mostly because land-based policies can only be adopted by nation states domestically and require addressing major systemic issues such as waste management and disposal of plastic within national territory. Environmentally sound management practices that are cost-effective are still being developed and so effectively addressing plastic at source has been challenging. Sea-based sources on the other hand, such as fishing gear and discharge from vessels, were relatively easily addressed in the MARPOL Convention as early as 1988 as it did not require costly mechanisms to address these. Hence, the first UNEA resolution asked for further reports on the identification of the key sources, more research on impact — to gauge the scale of remedial action — and measures to address the issue.

By the second UNEA resolution in 2016, there was a greater

understanding of potential harm so the effort was directed toward identifying the different agencies e.g., the International Maritime Organization (IMO) and the Food and Agriculture Organization (FAO) that are involved in addressing this issue, bringing together additional stakeholders such as civil society and industry, and how national and regional mechanisms can be developed (80). By the third and fourth resolutions, it was clearer that national measures are important. However, very slow progress had been made on that front. Hence, the focus shifted to raising awareness about the impact, collecting data to establish urgency, and comparing different cost implications for potential measures — especially for developing countries — so that their concerns can be addressed. Moreover, there was also a realization that there are scattered agencies with different mandates and hence inter-stakeholder coordination is important to develop synergy between multiple prongs of action towards a common end.

Therefore, clearly, national, and regional legislative frameworks and plans are crucial to the effective addressal of this issue. Even the new plastic treaty will be unable to make any significant impact unless states are swift to act in curbing the plastic menace at a war footing. Therefore, greater attention needs to be place on operationalizing regional strategies at the national level.

National measures

As part of its endeavor to support state action on adopting national legislation, the UNEP released a toolkit for national legislation (81). The key findings focus on addressing single-use items and developing an inter-agency mechanism to co-ordinate the development, implementation, and review of national plans and strategies. However, mere legislation without active implementation and monitoring strategy is likely to be ineffective. Such strategies need to establish baseline conditions, set targets, and develop market-based incentives for the collection of plastic waste to enable effective review.

For example, India does have the Plastic Waste Management (Amendment) Rules 2021 under the Environment (Protection) Act 1986 to specifically deal with plastic waste management (82). It has banned the manufacture, import, stocking, distribution, sale and use of plastic carry bags of thickness less than 120 microns and 19 single-use plastic products. Additionally, the rules even prescribe the colors and pigments that may be used in carrying plastic bags (even if only in the context of those plastic bags in contact with food, pharmaceuticals, and drinking water). This list should be updated to account for the way marine biodiversity interacts with these colors which eventually enter the food chain. The Government of India has also established a Special Task Force and National Level Task Force to implement and coordinate these rules (83). In addition to an incomprehensive addressal of microplastics — the textile and tyre industry has not been addressed. Four years later, there is poor implementation of these norms (84). This is primarily because of: a) the lack of an action/impl ementation plan; b) the lack of viable alternatives; c) poor compliance from manufactures due to effective lobbying; d) poor union state coordination; and e) lack of behavioral change on plastic consumption. All this combined leads to poor compliance, monitoring, and enforcement.

This also highlights the need for all stakeholders from raw material producers, manufacturers, state officials, and end-consumers to ensure the success of any strategy. While there is a distinction between consumer plastic litter and manufacturer plastic litter, environmentally conscious consumption patterns will have an impact on business practices too. Therefore, national strategies also need to focus on public advocacy as reducing consumption of plastic goods can go a long way in reducing generation of plastic waste. Advocacy can also be useful in assisting the state to ensure proper waste management. Additionally, simply regulating this issue without supporting waste management and recycling infrastructure is unlikely to make a dent.

Concluding comments

Marine microplastic pollution represents a complex and intensifying global crisis that transcends ecological boundaries, threatening marine biodiversity, compromising food systems, and posing potential risks to human health. Its persistence reflects a cyclical challenge. Limitations in current detection methodologies hinder robust data collection, which in turn delays regulatory responses and perpetuates fragmented, reactive interventions. This complexity is further deepened by the heterogeneity of microplastics—their varying sizes, shapes, chemical additives, polymer types, and degrees of degradation—all of which influence environmental behavior, bioavailability, and trophic transfer.

An overemphasis on downstream technological solutions including AI and portable detection tools often detracts from addressing upstream systemic drivers such as unsustainable production models and inadequate waste governance. Simultaneously, the prevailing marine-centric focus frequently neglects terrestrial and atmospheric pathways, resulting in incomplete risk assessments and significant policy blind spots. Addressing these multifaceted challenges demands that standardization precedes innovation to ensure data consistency and comparability. Regulatory frameworks must adopt precautionary, life-cycle-based approaches rather than waiting for conclusive human health evidence to be obtained.

Without confronting these interlinked scientific, structural, and governance gaps, mitigation efforts will remain piece-meal, reactive, and ultimately insufficient—placing ecosystems and public health at continued risk. Escalating healthcare costs tied to potential microplastic exposure underscore the urgency of investing in long-term toxicity research. Meanwhile, projections indicating that global waste generation will exceed current management capacities highlight the need for stringent industrial regulations and corporate accountability across the plastic value chain.

A paradigm shift toward holistic, system-based solutions is essential. This includes integrating scientific innovation with regulatory foresight and societal engagement. Strengthening marine monitoring systems, improving waste infrastructure, and enhancing coordination across national, regional, and global governance mechanisms are critical to controlling microplastic proliferation. In addition to a legislative framework, a supporting implementing strategy with defined goals, clear institutional roles, and enforceable timelines is necessary to galvanize collective action and enable everyone to play their part.

As global plastic production continues to accelerate, the window for effective intervention is narrowing. Reactive, siloed responses are no longer sufficient. What is urgently needed is a cohesive, globally coordinated strategy—anchored in precaution, informed by robust science, and aligned with regional and national legal frameworks as it is imperative to safeguard ocean health, food security, and human wellbeing for current and future generations.

References

1.Fava M. Ocean plastic pollution an overview: data and statistics. UNESCO Intergovernmental Oceanographic Commission; 2022 May 9. https://oceanliteracy.unesco.org/plastic-pollution-ocean/

2.Jakubowicz I. Evaluation of degradability of biodegradable polyethylene (PE). Polym Degrad Stab. 2003;80(1):39–43. https://doi.org/10.1016/S0141-3910(02)00383-3

3.Barnes DKA, Galgani F, Thompson RC & Barlaz M. Accumulation and fragmentation of plastic debris in global environments. Philos Trans R Soc Lond B Biol Sci. 2009;364:1985–98. https://doi.org/10.1098/rstb.2008.0205

4.Moore CJ. Synthetic polymers in the marine environment:

A rapidly increasing, long-term threat. Environ Res. 2008;108(2):131–9. https://doi.org/10.1016/j.envres.2008.07.025

5.Ryan PG & Moloney CL. Marine litter keeps increasing. Nature. 1993;361:23. https://doi.org/10.1038/361023a0

6.Haward M. Plastic pollution of the world's seas and oceans as a contemporary challenge in ocean governance. Nat Commun. 2018;9:667. https://doi.org/10.1038/s41467-018-03104-3
7.Leal Filho W, Havea PH, Balogun A-L, Boenecke J, Maharaj AA, Ha'apio M, et al. Plastic debris on Pacific islands: ecological and health implications. Sci Total Environ. 2019;670:181–7. https://doi.org/10.1016/j.scitotenv.2019.03.184

8. Walkinshaw C, Lindeque PK, Thompson R, Tolhurst T & Cole M. Microplastics and seafood: lower trophic organisms at highest risk of contamination. Ecotoxicol Environ Saf. 2020;190:110066. https://doi.org/10.1016/j.ecoenv.2019.110066

9.Sivadas SK, Ramu K, Mishra P & Murthy RMV. Potential plastic accumulation zones in the Indian coastal seas. Front Mar Sci. 2021;8:768001. https://doi.org/10.3389/fmars.2021.768001

10. Thompson RC, Olsen Y, Mitchell RP, Davis A, Rowland SJ, John AWG, et al. Lost at sea: where is all the plastic? Science. 2004;304(5672):838. https://doi.org/10.1126/science.1094559

11. Moore CJ. Synthetic polymers in the marine environment: A rapidly increasing, long-term threat. Environ Res. 2008;108(2):131–9. https://www.sciencedirect.com/science/article/abs/pii/S0048969721069278

12. Lebreton L, Slat B, Ferrari F, Sainte-Rose B, Aitken J, Marthouse R, et al. Evidence that the Great Pacific Garbage Patch is rapidly accumulating plastic. Sci Rep. 2018;8(1):4666. https://doi.org/10.1038/s41598-018-22939-w

13. United Nations News. Turn the tide on plastic, urges UN, as microplastics in the seas now outnumber stars in our galaxy. United Nations. 2017 Feb. https://news.un.org/en/story/2017/02/552052-turn-tide-plastic-urges-un-microplastics-seas-now-outnumber-stars-our-galaxy

14. Schwabl P, Köppel M, Königshofer J, Bucsics A, Trauner A & Liebmann R. Assessment of microplastic contamination in human stool—a pilot study. Ann Intern Med. 2019;171(7):453–7. https://doi.org/10.7326/M19-0618

15. United Nations Environment Programme (UNEP). Plastic waste inputs from land into the ocean. In: Bergmann M, Gutow L, Klages M, editors. Marine anthropogenic litter. Cham: Springer; 2015; 273–302. https://link.springer.com/chapter/10.1007/978-3-319-16510-3_7

16. United Nations Environment Programme (UNEP). Plastic waste inputs from land into the ocean. In: Bergmann M, Gutow L, Klages M, editors. Marine anthropogenic litter. Cham: Springer; 2015; 273–302. https://link.springer.com/chapter/10.1007/978-3-319-16510-3_7

17. Boucher J & Friot D. Primary microplastics in the oceans: a global evaluation of sources. Gland (Switzerland): International Union for Conservation of Nature (IUCN); 2017. https://doi.org/10.2305/IUCN.CH.2017.01.en
Also See: European Parliament. Microplastics: sources, effects, and solutions. European Parliament, Europa; 2018. https://www.europarl.europa.eu/topics/en/article/20181116 STO19217/microplastics-sources-effects-and-solutions

18. Anderson PJ, Warrack AJ, Langen VL, Challis BS & Hanson GD. Microplastic contamination in Lake Winnipeg, Canada. Environ Pollut. 2017;225:223–31. https://doi.org/10.1016/j.en vpol.2017.02.072

19. Lebreton L, Slat B, Ferrari F, Sainte-Rose B, Aitken J, Marthouse R, et al. Evidence that the Great Pacific Garbage Patch is rapidly accumulating plastic. Sci Rep. 2018;8(1):4666. https://doi.org/10.1038/s41598-018-22939-w

20. Lebreton L, Slat B, Ferrari F, Sainte-Rose B, Aitken J, Marthouse R, et al. Evidence that the Great Pacific Garbage Patch is rapidly accumulating plastic. Sci Rep. 2018;8(1):4666. https://doi.org/10.1038/s41598-018-22939-w

21. Lebreton L, Slat B, Ferrari F, Sainte-Rose B, Aitken J, Marthouse R, et al. Evidence that the Great Pacific Garbage Patch is rapidly accumulating plastic. Sci Rep. 2018;8(1):4666. https://doi.org/10.1038/s41598-018-22939-w

22. Blue Growth. Index: Rivers and lakes of the world – A to Z listing. Blue Growth. 2025 Apr 5. https://www.blue-g

rowth.org/Oceans_Rivers_Seas/Index_Rivers_Lakes_Of_
The_World_%20A_To_Z_Listing.htm

23. Andrady AL. Microplastics in the marine environment. Mar Pollut Bull. 2011;62(8):1596–605. https://doi.org/10.1016/j.marpolbul.2011.05.030

24. Thompson RC, Napper IE & Galloway TS. The plastic paradox: Environmental trade-offs of plastic pollution and its mitigation. Nat Clim Change. 2020;10(5):415–21. https://doi.org/10.1038/s41558-020-0746-1

25. Galgani, François, Georg Hanke & Thomas Maes. *"Global distribution, composition and abundance of marine litter." Marine Pollution Bulletin.* 142 (2019): 1–8. https://doi.org/10.1016/j.marpolbul.2019.03.019.

26. Galgani F, Hanke G & Maes T. Global distribution, composition and abundance of marine litter. Mar Pollut Bull. 2019;142:1–8. https://doi.org/10.1016/j.marpolbul.2019.03.019

27. Mato Y, Isobe T, Takada H, Kanehiro H, Ohtake C, Kaminuma T. Plastic resin pellets as a transport medium for toxic chemicals in the marine environment. Environ Sci Technol. 2001;35(2):318–24. https://doi.org/10.1021/es0010498

28. Kenyon KW & Kridler E. Laysan Albatrosses swallow indigestible matter. Auk. 1969;86(4):744-5. https://digitalcommons.usf.edu/cgi/viewcontent.cgi?article=18893&context=auk

Also refer: Rothstein SI. Plastic particle pollution of the surface of the Atlantic Ocean: Evidence from a seabird. Condor. 1973;75(3):344-5. https://sora.unm.edu/sites/default/files/journals/condor/v075n03/p0344-p0345.pdf

29. Smith M, Love DC, Rochman CM & Neff RA. Microplastics in seafood and the implications for human health. Curr Environ Health Rep. 2018;5(3):375–86. https://doi.org/10.1007/s40572-018-0206-z

30. Rochman CM, Hoh E, Kurobe T, Teh SJ. Ingested plastic transfers hazardous chemicals to fish and induces hepatic stress. Sci Rep. 2013;3:3263. https://doi.org/10.1038/srep03263

31. Galloway TS, Cole M & Lewis C. Interactions of microplastic debris throughout the marine ecosystem. Nat Ecol Evol. 2017;1(5):0116. https://doi.org/10.1038/s41559-017-0116

32. Wright SL & Kelly FJ. Plastic and human health: a micro issue? Environ Sci Technol. 2017;51(12):6634–47. https://doi.org/10.1021/acs.est.7b00423

33. Rochman CM, Hoh E, Kurobe T & Teh SJ. Ingested plastic transfers hazardous chemicals to fish and induces hepatic stress. Sci Rep. 2013;3:3263. https://doi.org/10.1038/srep03263

34. Porter A, Godbold JA, Lewis C & Savage G. Microplastic burden in marine benthic invertebrates depends on species traits and feeding ecology within biogeographical provinces.

Nat Commun. 2023;14(1). https://doi.org/10.1038/s41467-02
3-43788-w

35. Rochman CM, Browne MA, Halpern BS, Hentschel BT, Hoh E, Karapanagioti HK, et al. Policy: Classify plastic waste as hazardous. Nature. 2013;494(7436):169–71. https://doi.org/10.1038/494169a

36. Wang J, Tan Z, Peng J, Qiu Q & Li M. The impact of microplastic pollution on the reproductive system: a critical review. Environ Pollut. 2021;269:116133. https://doi.org/10.1016/j.envpol.2020.116133

37. Li D, Shi Y, Yang L, Xiao L, Kehoe DK, Gun'ko YK & Wang J. Microplastic exposure and cancer risk: a critical review of current data and future perspectives. Environ Res. 2022;204:112073. https://doi.org/10.1016/j.envres.2021.11207 3

38. Prata JC, da Costa JP, Lopes I, Duarte AC & Rocha-Santos T. Environmental exposure to microplastics: an overview on possible human health effects. Sci Total Environ. 2020;702:134455. https://doi.org/10.1016/j.scitotenv.2019.13 4455

39. Barboza LGA, Vethaak AD, Lavorante BRBO, Lundebye AK & Guilhermino L. Marine microplastic debris: an emerging issue for food security, food safety, and human health. Mar Pollut Bull. 2018;133:336–48. https://doi.org/10.1016/j.marp olbul.2018.05.047

40. Gigault J, Pedrono B, Maxit B & Ter Halle A. Marine plastic litter: The unanalyzed nano-fraction. Environ Sci Nano. 2018;3(2):346–50. https://doi.org/10.1039/C5EN00191A

41. Rummel CD, Jahnke A, Gorokhova E, Kühnel D & Schmitt-Jansen M. Impacts of biofilm formation on the fate and potential effects of microplastic in the aquatic environment. Environ Sci Technol Lett. 2016;3(7):258–65. https://doi.org/10.1021/acs.estlett.6b00264

42. Barboza LGA, Cunha SC, Monteiro C, Fernandes JO & Guilhermino L. Bisphenol A and other plastic-associated chemicals in fish and shellfish: a risk to human health? J Hazard Mater. 2020;393:122419. https://doi.org/10.1016/j.jha zmat.2020.122419

43. Cox KD, Covernton GA, Davies HL, Dower JF, Juanes F & Dudas SE. Human consumption of microplastics. Environ Sci Technol. 2019;53(12):7068–74. https://doi.org/10.1021/acs.est.9b01517

44. Ory NC, Sobral P, Ferreira JL & Thiel M. Amberstripe scad Decapterus muroadsi as a bioindicator for microplastic pollution in the central Pacific Ocean. Environ Pollut. 2018;244:566–73. https://doi.org/10.1016/j.envpol.2018.10.013

45. Savoca MS, Wohlfeil ME, Ebeler SE & Nevitt GA. Marine plastic debris emits a keystone infochemical for olfactory foraging seabirds. Sci Adv. 2017;2(11):e1600395. https://doi.org/10.1126/sciadv.1600395

46. Ory NC, Sobral P, Ferreira JL & Thiel M. Amberstripe scad Decapterus muroadsi as a bioindicator for microplastic pollution in the central Pacific Ocean. Environ Pollut. 2018;244:566–73. https://doi.org/10.1016/j.envpol.2018.10.013

47. Rummel CD, Jahnke A, Gorokhova E, Kühnel D & Schmitt-Jansen M. Impacts of biofilm formation on the fate and potential effects of microplastic in the aquatic environment. Environ Sci Technol Lett. 2016;3(7):258–65. https://doi.org/10.1021/acs.estlett.6b00264

48. Watts AJR, Urbina MA, Corr S, Lewis C & Galloway TS. Ingestion of plastic microfibers by the crab Carcinus maenas and its effect on food consumption and energy balance. Environ Sci Technol. 2015;49(24):14597–604. https://doi.org/10.1021/acs.est.5b04026

49. Prata JC, Duarte AC & Rocha-Santos T. Methods for sampling and detection of microplastics in water and sediment: a critical review. Trends Analyt Chem. 2019;110:150–9. https://doi.org/10.1016/j.trac.2018.10.030

50. Rochman CM, Hoh E, Kurobe T & Teh SJ. Ingested plastic transfers hazardous chemicals to fish and induces hepatic stress. Sci Rep. 2019;9(1):1–7. https://doi.org/10.1038/s41598-019-42430-8

51. Tagg AS, Harrison PS, Ju-Nam S & Ojeda RJ. Microplastic detection methods and their limitations. Mar Pollut Bull. 2015;99(1–2):140–5. https://doi.org/10.1016/j.marpolbul.2

015.08.009

52. Käppler A, Windrich S, Löder M, et al. Identification of microplastics by FTIR and Raman microscopy: a novel silicon filter substrate opens the important spectral range below 1300 cm⁻¹ for FTIR transmission measurements. Anal Bioanal Chem. 2016;408(21):5831–41. https://doi.org/10.100 7/s00216-016-9598-9

53. Hidalgo-Ruz V, Gutow L, Thompson RC & Thiel M. Microplastics in the marine environment: a review of the methods used for identification and quantification. Environ Sci Technol. 2012;46(6):3060–75. https://doi.org/10.1021/es203 1505

54. Hidalgo-Ruz V, Gutow L, Thompson RC & Thiel M. Microplastics in the marine environment: a review of the methods used for identification and quantification. Environ Sci Technol. 2012;46(6):3060–75. https://doi.org/10.1021/es203 1505

55. Koelmans AA, Kooi M, Law KL & van Sebille E. All is not lost: Deriving a top-down mass budget of plastic at sea. Environ Res Lett. 2019;14(11):114028. https://iopscience.iop.org/article/ 10.1088/1748-9326/aa9500

56. Free CM, Jensen OP, Mason SA, Eriksen M, Williamson NJ & Boldgiv B. High levels of microplastic contamination in a large, remote, mountain lake. Mar Pollut Bull. 2014;85(1):156–63. https://pubmed.ncbi.nlm.nih.gov/24973278/

57. Woodall LC, Sanchez-Vidal A, Canals M, Paterson GL,

Coppock R & Sleight V, Thompson RC. The deep sea is a major sink for microplastic debris. R Soc Open Sci. 2015;2(5):140317. https://doi.org/10.1098/rsos.140317

58. Lei K, Chen SQ, Xu B, Li Y & Wang LL. Microplastic identification and quantification using deep learning: a new approach for environmental monitoring. Sci Total Environ. 2021;778:146164. https://doi.org/10.1016/j.scitotenv.2021.14 6164

59. Murray G. Environmental implications of plastic debris in marine settings – entanglement, ingestion, smothering, hangers-on, hitch-hiking and alien invasions. Philos Trans R Soc Lond B Biol Sci. 2009 Jul 27;364(1526):2013–25. https://d oi.org/10.1098/rstb.2008.0265

60. European Commission. A European strategy for plastics in a circular economy. COM (2018) 28. 2018 Jan 16. https://eur-lex.europa.eu/resource.html?uri=cellar:2df5d1d2-fac7-11e7-b8f5-01aa75ed71a1.0001.02/DOC_1&format=PDF

61. European Commission. A European strategy for plastics in a circular economy. COM (2018) 28. 2018 Jan 16. https://eur-lex.europa.eu/resource.html?uri=cellar:2df5d1d2-fac7-11e7-b8f5-01aa75ed71a1.0001.02/DOC_1&format=PDF

62. European Commission. Pathway to a healthy planet for all, EU action plan: 'Towards zero pollution for air, water and soil'. COM (2021) 400. 2021 May 12. https://eur-lex.eur opa.eu/resource.html?uri=cellar:a1c34a56-b314-11eb-8aca-01aa75ed71a1.0001.02/DOC_1&format=PDF

63. European Union. Regulation (EC) No 1907/2006 of the European Parliament and of the Council concerning the Registration, Evaluation, Authorisation and Restriction of Chemicals (REACH). European Union. 2006. https://eur-lex. europa.eu/legal-content/EN/TXT/?uri=CELEX:32006R1907

64. European Union. Commission Regulation (EU) 2023/2055 amending Annex XVII to Regulation (EC) No 1907/2006 of the European Parliament and of the Council concerning the Registration, Evaluation, Authorisation and Restriction of Chemicals (REACH) as regards synthetic polymer microparticles. Off J Eur Union. 2023 Sep 25. https://eur-lex.europa.eu/legal-content/EN/TXT/PDF/?uri=CELEX:32023R2055

65. European Union. Commission Regulation (EU) 2023/2055 amending Annex XVII to Regulation (EC) No 1907/2006 of the European Parliament and of the Council concerning the Registration, Evaluation, Authorisation and Restriction of Chemicals (REACH) as regards synthetic polymer microparticles. Off J Eur Union. 2023 Sep 25. https://eur-lex.europa.eu/legal-content/EN/TXT/PDF/?uri=CELEX:32023R2055

66. European Union. Regulation (EC) No 1907/2006 of the European Parliament and of the Council concerning the Registration, Evaluation, Authorisation and Restriction of Chemicals (REACH). European Union. 2006. https://eur-lex. europa.eu/legal-content/EN/TXT/?uri=CELEX:32006R1907

67. European Union. Regulation (EC) No 1907/2006 of the European Parliament and of the Council concerning the Registration, Evaluation, Authorisation and Restriction of Chemicals (REACH). European Union. 2006. https://eur-lex.

europa.eu/legal-content/EN/TXT/?uri=CELEX:32006R1907

68. European Environment Agency. Microplastics from textiles: towards a circular economy for textiles in Europe. EEA Europa; 2023 Feb 10. https://www.eea.europa.eu/publica tions/microplastics-from-textiles-towards-a

69. European Union. Explanatory guide to REACH restriction of synthetic polymer microparticles. European Union. 2025. https://webgate.ec.europa.eu/circabc-ewpp/d/d/workspace/ SpacesStore/a4b3c599-db77-4210-8ca1-430e88c59bb1/file. bin

70. Bruch C, et al. Marine litter legislation: a toolkit for poli-cymakers. Nairobi: United Nations Environment Programme. 2016. https://www.unep.org/resources/report/marine-litter -legislation-toolkit-policymakers

71. Thompson RC. Microplastics in the marine environment: sources, consequences and solutions. In: Bergmann M, Gutow L, Klages M, editors. Marine anthropogenic litter. Cham: Springer International Publishing; 2015; 185–200. https://do i.org/10.1007/978-3-319-16510-3_7

72. European Union. Commission Regulation (EU) 2023/2055 amending Annex XVII to Regulation (EC) No 1907/2006 of the European Parliament and of the Council concerning the Registration, Evaluation, Authorisation and Restriction of Chemicals (REACH) as regards synthetic polymer microparti-cles. Off J Eur Union. 2023 Sep 25. https://eur-lex.europa.eu/ legal-content/EN/TXT/PDF/?uri=CELEX:32023R2055

73. European Union. Commission Regulation (EU) 2023/2055 amending Annex XVII to Regulation (EC) No 1907/2006 of the European Parliament and of the Council concerning the Registration, Evaluation, Authorisation and Restriction of Chemicals (REACH) as regards synthetic polymer microparticles. Off J Eur Union. 2023 Sep 27;L 238:67–88. https://eur-lex.europa.eu/legal-content/EN/TXT/PDF/?uri=CELEX:3202 3R2055

74. European Union. Explanatory Guide to REACH restriction of synthetic polymer microparticles. European Union.

75. United States Congress. Microbead-free waters act of 2015, H.R.1321, 114th Congress (2015–2016). United Nations. https://www.congress.gov/bill/114th-congress/house-bill/1 321/text

76. Lapyote Prasittisipin et al. Microplastics in construction and built environment. Developments in the Built Environment. 2023 Oct. https://www.sciencedirect.com/science/arti cle/pii/S2666165923000704#bib165

77. United Nations Environment Programme. Draft report of the intergovernmental negotiating committee to develop an international legally binding instrument on plastic pollution, including in the marine environment, on the work of the first part of its fifth session. UNEP/PP/INC.5/8. 2025 Feb 10. https://wedocs.unep.org/bitstream/handle/20.500.11822/47 162/INC_5_1_Report.pdf

78. United Nations Environment Programme. UNEA Res-

olution 5/14 entitled "End plastic pollution: Towards an international legally binding instrument". UNEP/PP/OEWG/1/INF/1. 2022 May 10. https://wedocs.unep.org/bitstream/handle/20.500.11822/39812/OEWG_PP_1_INF_1_UNEA%20resolution.pdf

79. United Nations Environment Programme. Compilation of United Nations Environment Assembly resolutions on marine litter and microplastics. UNEP/AHEG/2019/3/INF/2. 2019 Oct 25. https://apps1.unep.org/resolutions/uploads/uneaml_en.pdf

80. United Nations Environment Programme. Compilation of United Nations Environment Assembly resolutions on marine litter and microplastics. UNEP/AHEG/2019/3/INF/2. 2019 Oct 25. https://apps1.unep.org/resolutions/uploads/uneaml_en.pdf

81. United Nations Environment Programme. Marine litter legislation: A toolkit for policymakers. UNEP. 2016 May 9. https://www.unep.org/resources/report/marine-litter-legislation-toolkit-policymakers

82. Ministry of Environment, Forest and Climate Change, Government of India. Plastic waste management (amendment) rules, 2018. Central Pollution Control Board. 2018 Mar 27. https://cpcb.nic.in/displaypdf.php?id=cGxhc3RpY3dhc3RlL1BXTV9HYXpldHRlLnBkZg==

83. Ministry of Environment & Forest and Climate Change. Ban on single use plastics. Press Information Bureau. 2022 Dec

12. https://pib.gov.in/PressReleasePage.aspx?PRID=1882855

84. Krishnan M. Why is India's single-use plastic ban failing? DW News. 2022 Feb 11. https://www.dw.com/en/why-is-indias-single-use-plastic-ban-failing/a-63625217

5

Coastal Water Dynamics and Human Health: Vulnerabilities and Resilience

Abstract

Coastal water systems are increasingly threatened by anthropogenic pollution and climate-driven alterations, creating
significant challenges for human health and coastal ecosystem
integrity. This chapter explores the multifaceted relationship
between coastal water dynamics, pollution patterns, and associated health outcomes in coastal communities. Industrial
contaminants, including heavy metals from activities like
mining, introduce toxins such as lead and arsenic into groundwater systems that eventually discharge into marine envi-
ronments.These pollutants have numerous health impacts
including salmonellosis, typhoid fever, hepatitis, cholera, and
dysentery through direct contact or by the consumption of
contaminated seafood associated with bioaccumulation of
marine toxins.

The authors of this chapter argue that coastal water quality requires coordinated action across multiple sectors, with particular attention to vulnerable populations in coastal areas where monitoring and infrastructure remain inadequate. A PRISMA review is conducted to analyze and explore the intricate relationship between coastal water pollution, warming waters, and human health. Management and governance approaches, including water quality monitoring and wastewater treatment technologies, as well as adaptation strategies for mitigating these health risks are examined.

Keywords: Coastal water, vulnerabilities, anthropogenic, pollution, human health, coastal ecosystem, heavy metals.

Authors

Gulshan Sharma, Associate Fellow, Resilience and Sustainability of Ocean Resources (RSOR) Cluster, National Maritime Foundation, India

Tariq Ahmad, Junior Research Associate (JRA) (RSOR Cluster), National Maritime Foundation (NMF), India

Introduction

The intricate relationships between coastal marine ecosystems and human health represent one of the most critical yet understudied areas of environmental health science. These dynamic interfaces between land and sea not only support extraordinary biodiversity but also provide essential services that sustain human communities worldwide. As anthropogenic pressures intensify through climate change, pollution, and habitat destruction, the implications for both ecosystem integrity and human wellbeing have become increasingly pronounced.

Importance of coastal water ecosystems: Ecological and economic roles of marine environments

Coastal and marine ecosystems represent some of the most productive and diverse environments on Earth, providing a range of essential services that support both ecological processes and human activities. Research on marine and coastal ecosystem services (MCES) has grown exponentially in recent decades, revealing the critical importance of these environments for global wellbeing. A comprehensive systematic review identified 476 indicators for assessing MCES, highlighting the complexity and breadth of benefits derived from these systems (1). Food provision, particularly through fisheries, remains the most extensively analyzed service, reflecting the crucial role of marine environments on global food security.

The economic value of these ecosystem services is substantial with estimates suggesting that marine and coastal ecosystems contribute trillions of dollars annually to the global economy through direct and indirect benefits. Mangroves and coastal wetlands emerge as particularly important habitats, providing key functions for marine ecosystems and substantial economic benefits. Unfortunately, these habitats are disappearing at alarming rates—35 percent of mangroves have been lost in recent decades, while some regions report annual coastal wetland losses of up to 20 percent. Such losses threaten not only biodiversity but also the livelihoods and wellbeing of hundreds of millions of people worldwide (2).

Coral reefs represent another critical coastal ecosystem, providing habitat for commercially important fish species while offering coastal protection from storms and erosion (3). Similarly, seagrass meadows deliver essential services through their roles in carbon storage, erosion control, and as nursery habitats for marine species. The diversity of services provided by these various coastal ecosystems underscores their irreplaceable value to human societies globally.

Interconnectedness of marine and human health: Exploring the complex relationships between marine ecosystems and human wellbeing

The relationship between marine environments and human health encompasses multidimensional connections that influence physical, mental, and social wellbeing. Contrary to common perceptions that associate health inequalities primarily with inner-city environments, coastal communities often experience significant health challenges that have received insufficient attention. Recent mapping of key health indicators reveals a distinct core-periphery pattern in disease prevalence, with coastal communities bearing a disproportionately high burden of ill health across multiple conditions (4).

Marine environments affect human health through numerous pathways including nutrition, exposure to pathogens and toxins, natural disaster mitigation, and opportunities for recreation and psychological restoration (5). The quality of coastal waters directly influences human health through exposure during recreational activities and consumption of seafood. Harmful algal blooms, bacterial contamination, and chemical pollutants can lead to acute illnesses and chronic health conditions among coastal populations (6)(7). Conversely, access to healthy marine environments provides opportunities for physical activity, stress reduction, and improved mental health outcomes (8).

Of particular concern are the emerging patterns of health outcomes for children and young people in coastal areas. These patterns potentially reflect a shift in the distribution of child poverty since the 1990s and may signal an impending public health crisis in coastal communities without appropriate intervention.

Materials and methods

We followed a systematic review method to report the association between coastal water dynamics and human health and related disease. The research was conducted in consent with the criteria of PRISMA (Preferred Reporting Items for Systematic Reviews and Meta-Analyses). The flow diagram is shown in Fig. 1. The research was guided in the Scopus database using Advanced Search Builder and the keywords were searched in (Abstract or Title). We filtered the review and research articles published only in the English language and selected the keywords: marine pollution, water quality, coastal ecosystems, public health, climate change, eutrophication. We excluded articles, short surveys, letter, opinions, commentary, or non-relevant articles. We obtained a total of 62 appropriate published articles in their final version. For some of the papers, we selected only principal findings that precisely fit the aim of this review.

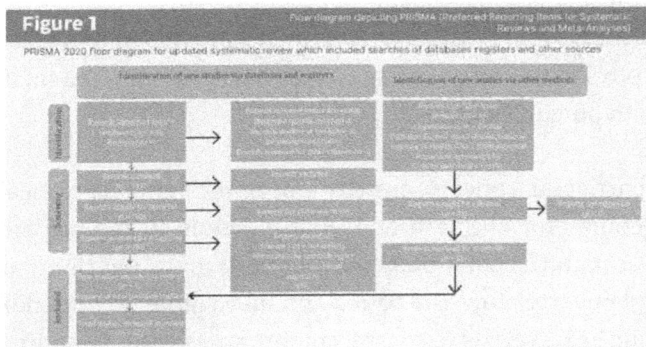

Source: BMJ 2021 (https://www.prisma-statement.org/prisma-2020-flow-diagram)

Coastal water pollution sources and impacts

Degradation of water resources is a major challenge across the world, with a critical impact on human health, food security, sanitation and hygiene as a first order impact. Coastal regions as an interface between the oceans — a foundational media for major biogeochemical cycles— and coastal lands which are inhabited by a major portion of the population become extremely vulnerable to impacts of water pollution (9). Water pollution in a coastal settlement impacts not only the health of its inhabitants but also poses risk to local oceanic fauna which might be the major food or protein source for the settlement. Furthermore, the ocean — being a major conduit for matter and energy — circulating resources across the world, the same

174

pollutants can have a much wider spatial range of impact (10).

Anthropogenic sources contribute to water pollution in the most significant manner and are much more feasible to mitigate in theory. Geogenic sources also exist, which can alter the composition of various chemicals in the water above the safe prescribed limits (11) . Geogenic pollution can pose a problem ranging from minor treatment methods for certain uses (drinking) to absolute prohibition on water usage and intensive treatment, if possible, as in case of arsenic contamination (12). This chapter focuses on anthropogenic sources of water pollution.

Water pollution (anthropogenic) can be understood in terms of type of pollutants – municipal sewage and solid waste, agricultural, industrial, thermal, etc. to name a few. It can be also classified in terms of the nature of discharge – point sources, which as the name suggests are sources which can be identified to be originating from a point, like a drain opening or a landfill. This is contrasted by non-point sources, which can't be attributed to a single point, but rather are a spatially distributed phenomenon like a surface runoff or percolation from an already contaminated area like agricultural fields with intensive farming methods. Hence, in the case of a non-point source, one must address a practice (e.g. use of chemical fertilizers) rather than any specific point (13). Lastly, water pollution must also be understood from the perspective of the type of water being impacted. Typically, surface water which includes rivers, canals, lakes, ponds, etc. is the most vulnerable water sources as it is exposed to all types of pollution through many pathways. Surface runoffs, contaminated precipitation (e.g, acid rain), contamination through soil as well as anthropogenic discharge into these sources makes them most vulnerable. Furthermore, due to the general nature of these systems, which ensure a certain degree of interconnectivity with each other as well as with groundwater and ocean water, contamination of these sources can leak into any downstream water body as well. Groundwater, which in India, tends to be a major water source even in the coastal region is a slowly replenishing resource, which can be treated as non-renewable in certain cases within human timeframes (14). Coastal groundwater is vulnerable to both anthropogenic contamination and over exploitation along with seawater contamination affecting salinity and chemical composition (15). Ocean water, which as mentioned previously, is a globally connected entity and hence can be potentially susceptible to being polluted from a geographically farther area. However, the ocean has its own complex mechanism of mixing materials and energy and the situation can be exacerbated or mitigated depending on prevalent currents and other conditions (16). From the perspective of an ocean settlement though, problems can be more localized. Firstly, it will affect the flora and fauna of the immediate coastal area, threatening local food security. The impact of contamination can also negatively impact human health. Biomagnification can be seen as such an impact. Finally, degradation of coastal waters can further affect economic activities like tourism which in turn can threaten the local economy (17). These issues are mere first order effects and several externalities have been observed due to poor coastal water health.

The chapter explores major water pollution sources in coastal settlements to highlight major areas which need to be addressed. Some issues are a function of increasing population in such settlements and modern consumption practices like reliance on plastics.

Municipal waste

Municipal waste refers to waste collected by municipal entities like local governing bodies. It can exist in the form of municipal sewage and stormwater or as municipal solid waste, which is colloquially understood as 'garbage'.

Municipal wastewater, which can refer to wastewater collected from sewage and in several cases to stormwater as well. In piped sewerage systems (as opposed to in situ systems like discharge pits), wastewater, consisting of blackwater (discharge from kitchen and toilets) and greywater (discharge from other domestic and non-industrial use not described as blackwater) form major hazardous components of wastewater. Stormwater, which originates from the precipitation runoff over the surfaces of settlements, can be discharged through a separate system even allowing storage and groundwater recharge. In that case surface contamination and sediment flow are the major water pollution challenges. However, in several cases, the practice is to have a single system, used to carry both wastewater and stormwater, rendering them essentially the same in terms of pollution management (18). It is to be noted in various cases; this practice is a design necessity and not an oversight.

Municipal wastewater has been classified as a single biggest contributor to coastal water pollution from land-based activities in India (19). It comprises biological contaminants like faecal matter, nutrient and food particulates and other organic matter providing fertile ground for pathogen growth, which are also a significant component of the discharge. Pathogens like E. Coli can thrive in such environments and lead to waterborne diseases. It also includes chemical pollutants from household items like detergents, disinfectants and other cleaning agents along with medicine compounds released as part of human waste. Heavy metals and toxic compounds from unintended sources like poor waste disposal practices are a significant threat in India, where hazardous waste disposal practices are not implemented rigorously at household level.

Municipal solid waste (MSW) has been found to have limited impact on coastal water pollution (20). However, it can nevertheless affect freshwater sources both on the surface and groundwater. MSW is typically collected at the household level by the local governing body and stored at location, from where it can be eventually sent to a long-term storage at a landfill or destroyed via methods like incineration, after any processing and recycling (if done). It is the landfill which in the form of leachate, especially when managed improperly, leads to contamination of water sources. Typical composition of MSW leachate can include dissolved organic matter (DOM) along with suspended particulates, inorganic compounds, ammonia, heavy metals, etc. While the impact of leachate is generally localized, surface runoff and groundwater movements can transport contaminants much farther.

Industrial discharges

Industrial discharges are the most 'popular' threats to water sources and their impacts have been well documented in cases of pollution borne diseases like Minamata in Japan. In India, primarily heavy metals and other industrial chemicals like acids, bases, etc. can be found in linked water bodies (21)(22). The practice in India is to prohibit direct release of effluents into water bodies. A common effluent treatment plant (CETP) is used in most cases, especially when it comes to releasing wastewater into coastal waters (23). However, literature indicates that upstream contamination along the drainage of the rivers systems leading into coastal waters can still carry those effluents into coastal waters (24).

Industrial effluents in coastal waters can significantly change the local coastal ecosystem, leading to prominence of organisms which can survive the presence of the contaminants (25). Phytoplankton blooms near the local coast can severely degrade and alter the local ecosystem affecting food supply (26). Furthermore, the presence of heavy metals and other effluents susceptible to biomagnification can contaminate the upstream food supply. Local contamination of freshwater surface and ground sources due to effluent discharge remains a traditionally understood risk of industrial activities.

Agricultural discharge

Contemporary agricultural practices are a significant and diverse source of ecological degradation. Due to the nature of the practices, water resource degradation, both in terms of overexploitation and pollution, is a serious threat (27). Agricultural activities contribute to water pollution across a spectrum of dimensions, some more obvious than the others.

The most obvious and well-known impact is agricultural runoff.Pesticides and fertilizers runoff (chemical ones) is a well-documented and popularly known challenge of intensive farming. Organochlorine based pesticides like dichlorodiph enyltrichloroethane (DDT), hexachlorocyclohexane (HCH), Endosulfan, etc. are banned or regulated within the country, even though studies have indicated the presence of these compounds exceeds the safe limits defined by international and domestic bodies (28). It should be noted that several pesticides are allowed as malaria control measures and can still be found in the water supply. A lot of contamination is from residual pesticides left before the bans, which high-lights the challenge posed. Most of these compounds are persistent, bio accumulative, and toxic leading to a myriad of health effects including development disorders, cancers, immunosuppression, hormone disruption, and reproductive distortion (29). The health impacts are both chronic and acute depending on exposure. Pesticide residue is not limited to the water supply as it also enters the food chain as a residual entity on food products that are used.

Chemical fertilizers have significant first and second order impacts on water resources and human health. Industrial processes required in manufacture also add harmful sub-stances like arsenic, cadmium, lead, uranium, etc. as a by-product of the process (30). Nitrates and calcium can act as a wide range of irritants that can cause renal damage and negatively impact bone structure. The second order effect of overabundance of plant macronutrients from surface runoff into local drainage basins can lead to algal blooms in water reservoirs used for human or livestock use. The algal bloom outcompetes pre-existing flora and fauna within the water body leading to oxygen, sunlight, and nutrient deprivation which, in turn, leads to death and decay of those organisms leading to pathogenic and chemical contamination. Such water might end up being unfit for use and along with pesticide and fertilizer runoff become hazardous for most purposes.

Another significant impact of agriculture comes from animal waste, specifically manure, which is a fertile ground for various pathogens like coliforms, salmonella, etc. (31). Improperly isolated discharge pits for human faecal waste disposal can also leach pathogens in the surrounding soil and can have similar effects. Assessment of upstream sources of such contaminants combined with analysis of local drainage patterns, must be done to mitigate this issue.

Lastly, antibiotic usage in livestock feed and other prophylactic methods lead to discharge of the same in the soil and water supply resulting in the rapid emergence of antibiotic-resistant strains of various diseases (32). India consumes three percent of global antimicrobials used for food animals, which includes aquaculture as well (33).

Climate change effects

Sea level rise and ocean warming: Impacts on marine ecosystems and water quality

Climate change exerts a profound influence on marine ecosystems through multiple pathways with ocean warming and sea level rise being among the most significant. Unlike atmospheric temperatures that fluctuate considerably, ocean temperatures show steady and consistent warming trends. Recent research indicates that the rate of ocean warming has doubled over the past 20 years, with 2023 recording one of the highest increases since the 1950s. While the Paris Agreement aimed to limit global warming to below 2°C above pre-industrial levels, ocean temperatures have already increased by an average of 1.45°C globally, with hotspots exceeding 2°C in the Mediterranean, Tropical Atlantic, and Southern Oceans (34).

This accelerated warming drives numerous ecological changes including shifts in species distributions, altered phenology (timing of life cycle events), increased disease prevalence, and reduced productivity in some regions. Marine heatwaves— extended periods of anomalously high ocean temperatures— have increased in frequency and intensity causing mass mortality events and potentially irreversible ecosystem changes. These temperature anomalies threaten thermally sensitive organisms like corals, leading to mass bleaching events that compound existing pressures on reef ecosystems (35).

Sea level rise represents another critical climate change impact with global sea levels rising at accelerating rates. Ocean warming contributes approximately 40 percent to global sea level rise through thermal expansion with the rate of rise doubling over the past 30 years to reach a total of nine cms. This seemingly modest increase has already enhanced coastal flooding risks in many regions with significant implications for coastal infrastructure, freshwater supplies, and human settlements. Rising seas infiltrate coastal aquifers with salt-water compromising drinking water supplies and agricultural productivity in coastal zones worldwide (36).

Extreme weather events: Increased pollution, ocean acidification

Climate change intensifies the hydrological cycle, leading to more frequent and severe extreme weather events that disproportionately affect coastal regions. Hurricanes, typhoons, and cyclones draw energy from warming ocean surfaces, potentially increasing in intensity while delivering more precipitation per event. These storms mobilize pollutants through flooding, combined sewer overflows, and damage to industrial facilities and waste management infrastructure, creating acute pollution pulses in coastal waters (37).

Beyond storms, changing precipitation patterns alter the timing and magnitude of terrestrial runoff, affecting the delivery of nutrients, sediments, and contaminants to coastal environments (38). In regions experiencing more intense rainfall, increased erosion mobilizes soil-bound contaminants and overwhelms wastewater treatment systems leading to more frequent contamination events. Conversely, drought conditions concentrate pollutants in reduced water volumes while potentially increasing reliance on contaminated water sources when preferred supplies become limited.

Ocean acidification represents a direct chemical consequence of increased atmospheric carbon dioxide (39). With 25-30 percent of fossil fuel emissions absorbed by oceans, the resulting chemical changes reduce seawater pH and carbonate ion availability. These changes threaten calcifying organisms including commercially important shellfish and coral reef builders, while potentially altering the bioavailability and toxicity of certain pollutants. Since the 1960s, oceans have lost approximately two percent of their oxygen content due to warming temperatures and pollutants, with coastal areas experiencing particularly severe impacts. Research has identified roughly 500 coastal "dead zones" worldwide where depleted oxygen levels have eliminated almost all marine life (38).

Vulnerable populations and environmental challenges: Limited access to clean water and healthcare for vulnerable communities

Access to clean water and adequate healthcare represents a particular challenge for vulnerable coastal communities. Despite coastal locations, many communities struggle with freshwater scarcity due to saltwater intrusion, contamination, or inadequate infrastructure. Socioeconomic status strongly influences vulnerability to coastal environmental hazards with disadvantaged communities often occupying higher-risk locations and lacking resources for adaptation or relocation (40). Historical patterns of development and discrimination have frequently concentrated marginalized populations in areas most exposed to flooding, erosion, and industrial pollution. Climate change exacerbates these challenges through sea level rise, changing precipitation patterns, and increasing the frequency of extreme events that disrupt water supply and damage infrastructure (41).

Case studies of pollution-related health outbreaks

Examples from coastal communities: Documented health impacts

Numerous documented cases illustrate the health impacts of waterborne disease and environmental degradation. Waterborne diseases are a concern in areas where industrial operations, coastal development, and poor wastewater management lead to the pollution of marine and coastal waters. The primary causes of waterborne disease outbreaks in marine ecosystems are pollutants that introduce bacteria, viruses, and parasites through untreated sewage, industrial chemicals, and agricultural runoff (42)(43). These pathogens can enter human beings through the consumption of contaminated seafood and through exposure to contaminated water. These pathogens also affect marine organisms. The most common waterborne illnesses include cholera, hepatitis A, gastroenteritis, and typhoid fever which poses a major threat to public health by seriously jeopardizing it, especially in coastal communities which rely on fishing and tourism (44).

In Minamata, Japan, industrial discharge of methylmercury into coastal waters led to devastating neurological damage in local communities consuming contaminated seafood. This landmark case, identified in the 1950s, continues to provide lessons about bioaccumulation, delayed health effects, and the need for precautionary approaches to chemical management in coastal environments (45).

More recently, harmful algal blooms have triggered multiple health emergencies in coastal communities worldwide. Florida's recurrent red tide events cause respiratory irritation among beachgoers, economic losses in tourism and fishing industries, and potential neurological impacts from the consumption of contaminated shellfish. Research indicates that these events that have intensified in frequency and duration are potentially linked to nutrient pollution and climate change (38).

Figure 2

Marine pollution devastates ecosystems and human health: Minamata disease from mercury contamination in Japan highlights neurological damage; red tides caused by harmful algal blooms disrupt marine biodiversity and poison seafood; vibrio infections in the U.S. Gulf Coast increase due to warming waters; oil spills, like the Deepwater Horizon disaster, suffocate marine life and destroy livelihoods

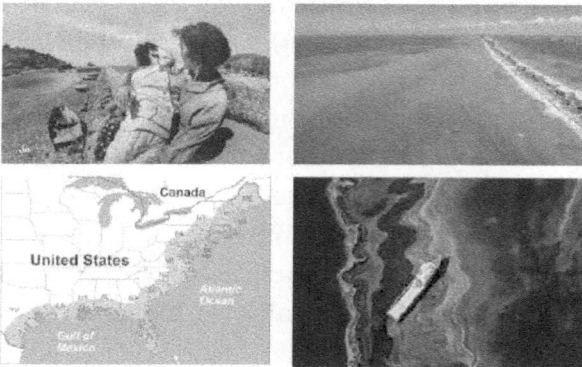

Source: The Japan Times. (2020, October 24). Minamata disease victims struggle to find closure. https://www.japantimes.co.jp/news/2020/10/24/national/social issues/minamata-disease-victims-closure/ Note: Additional images were obtained from Wikimedia Commons and are used in accordance with their respective Creative Commons licenses

Vibrio bacterial infections represent another emerging threat with warming coastal waters expanding the geographic range and seasonal windows for pathogenic species (44)(45). The

U.S. Gulf Coast has documented (Figure 2) increasing cases of wound infections and seafood-associated illness, particularly following extreme weather events that disrupt water and sanitation infrastructure (46).

Oil spills provide examples of acute pollution impacts on coastal community health (47). Studies following the Deepwater Horizon disaster documented respiratory symptoms, psychological distress, and endocrine disruption among cleanup workers and coastal residents. Long-term monitoring continues to reveal subtle impacts on community wellbeing including effects on mental health, social cohesion, and economic stability years after the initial event (48).

Management and governance approaches

Water resource management, which also covers the water quality and pollution aspects, is an important concept for the holistic management of water resources. The integrated water resource management (IWRM) approach is used to acquire more freshwater resources via rainwater harvesting and watershed management. Land use planning considering environmental conditions, is gaining traction to prevent contamination via both point and non-point sources. Governance plays a major role in the management of various water quality monitoring regimes.

Water pollution monitoring

In the United States of America, water pollution and safe drinking water is managed by the federal body, the US Environmental Protection Agency (EPA), empowered under the Clean Water Act (1972) and Safe Drinking Water Act (1974) along with various other legislations (49). Water quality standards are set by individual states. In contrast, in India, a single minimum water quality standard has been prescribed in BIS: 10500.

The Central Pollution Control Board (CPCB) is the apex body for monitoring pollution in India. CPCB along with the State Pollution Control Boards (SPCBs) are responsible for monitoring water quality across the country. These bodies maintain a water quality testing regime and are responsible for monitoring industrial discharges. However, at district level the District Water Sanitation Committee (DWSC) is responsible for monitoring and regulating the domestic water quality. It should be noted that CPCB and SPCB are under the Ministry of Environment, Forest and Climate Change as a statutory body, whereas DWSC is under the Ministry of *Jal Shakti*. The Water Quality Monitoring and Surveillance Framework (WQMS) was established under the National *Jal Jeevan* Mission (JJM) in 2019 (50). It aims to streamline the water quality monitoring regime especially for rural areas by empowering the Village Water/Sanitation Committee (VWSC) under the district counterpart (DWSC). JJM is focused on ensuring domestic water supply quality, whereas CPCB and SPCB still cover other domains maintaining their own network regime.

Water Quality Management Information System (WQMIS) compiles all the data provided by respective bodies under JJM, tracking quality and taking preventative measures for averting disease outbreaks (51).

Local governing bodies (urban and rural local bodies) are mandated to conduct water quality surveillance every month at the grassroot level, compiling the data into WQMIS. The JJM mandates monitoring drinking water at least once a year for chemical contamination and twice a year for biological contamination, setting a mandatory minimum surveillance requirement. Furthermore, it also specifies that no village can be left behind, providing a tool for financial support.

JJM also specifies standard operating procedure (SOP) for disease outbreaks, which must be framed by the respective State Water/ Sanitation Committee (SWSC). It also identifies water quality monitoring hotspots, specifically for chemical contamination either natural or anthropogenic. Any water supply infrastructure is scrutinized and approved by the local VWMS before development.

Wastewater treatment methods

Wastewater treatment involves removal of harmful substances and reconstitution of specified parameters below a prescribed criterion. Typically, it is desirable to recover water from waste through a sewage treatment plant (STP) and recirculate the water recovered. Eighty percent water can be recovered by this approach, with remaining lost along with the undesirable components. Regardless of the method used in between, a certain amount of wastewater must be released in a water body or disposed of on a drying bed. The only exception is bio septic tanks and their variants which aim to decompose organic waste, including faecal waste, locally and convert it into manure.

Treatment is conducted in stages with the primary stage being a physical process (52). This stage involves the physical removal of effluents via screening, grit, or sedimentation. This is followed by a secondary stage in which the sludge undergoes aerobic or anaerobic digestion by microbes with the difference being that in the aerobic system, oxygen containing air is pumped to biologically oxidize the waste. Various methods can be used to achieve the outcomes of these stages. Some of the methods include moving bed biofilm reactor (MBBR), sequencing batch reactor (SBR), up flow anaerobic sludge blanket (UASB) to name a few. The tertiary stage involves chemical and/or membrane technologies to remove phosphorus, heavy metals, and other effluents which can't be removed by previous methods. After this stage, the water still needs to be disinfected, which can be done via chlorination.

Newer methods based on electrochemical oxidation and photo-electrochemical oxidation are non-biological methods to achieve the objectives of the secondary step (53).

While the above process is required for safe drinking water, for other uses like flushing, gardening, or any other non-consumption use, water can be treated with alternative methods (54). Phytoremediation is one such method which can be used on grey water or in some cases on black water (55). It is based on the property of certain plants and their soil ecosystems to either absorb toxic metals within their body and concentrate them or decompose organic effluents via the microbiome around their roots. Recalcitrant pollutants, including organic ones can be removed from the soil and water (in the case of wetlands and on riparian buffers). Significant progress has been observed in the treatment of the *Ganga* River water in India using water hyacinths and certain algal species (56). Biochar, which is a charcoal generated from biomass, has also been suggested to have water purification capabilities.

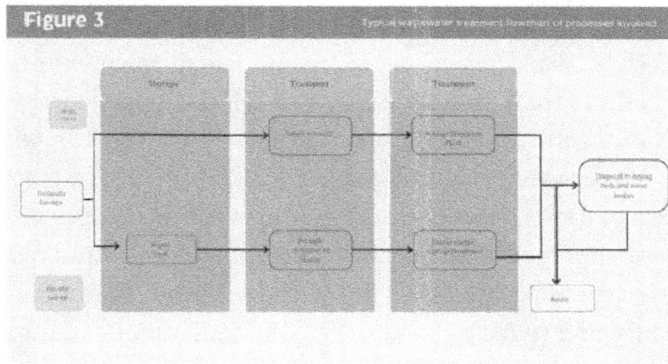

Source: Urban Wastewater Scenario in India, NITI Aayog, 2022 (54)

Adaptation strategies

Strategies for the reduction of marine water pollution require a holistic approach including community involvement, international cooperation, policy enforcement, and technological innovation. There is a need for tighter regulations on industrial discharges and agricultural practices to control chemical and nutrient pollution. Sustainable agricultural practices prevent nutrients from water bodies from running off to cause eutrophication. Included among such practices are organic farming, among others, and trying to minimize the use of chemical fertilizers. Response technologies for oil spills, safety regulations for marine transportation, offshore drilling, and the prevention of oil spills form an important part with all these initiatives. Initiatives such as The Ocean Cleanup are being initiated, as are advances in wastewater treatment to improve the removal and capture of pollutants before they reach the marine ecosystems.

Effective adaptation strategies include enhancing monitoring systems to track pollution trends and HAB occurrences using advanced nuclear and isotopic techniques. These methods allow for the precise identification of pollutant sources and temporal patterns enabling targeted interventions. Ecosystem-based approaches such as restoring mangroves, coral reefs, and seagrass beds, offer "soft protection" by improving water quality and reducing exposure to contaminants. Additionally, public health measures like early warning systems for HABs, seafood safety protocols, and community education campaigns can minimize health impacts. Infrastructure adaptations such as flood-proofing buildings and raising structures in vulnerable areas further reduce exposure risks. Collaborative efforts between governments, research institutions, and international organizations are essential to implementing these strategies effectively. By integrating scientific advancements with local actions, coastal zones can build resilience against marine pollution-related health outbreaks while promoting sustainable development.

Future directions and research priorities

Future directions must emphasize interdisciplinary collaboration, equitable governance, and innovative technological solutions. Emerging tools such as remote sensing and artificial intelligence can enhance real-time monitoring of pollutants like plastics, nutrient runoff, and oil spills, enabling rapid responses to contamination events. Research must address knowledge gaps in the cumulative effects of multiple stressors—such as ocean acidification, warming, and pollutant interactions—on marine ecosystems and human health, particularly in developing nations where coastal populations are projected to double in the upcoming years. Transdisciplinary approaches that integrate local ecological knowledge with scientific data are critical, as highlighted by disparities in priorities among scientists, policy-makers, and coastal communities. For instance, while scientists prioritize ocean acidification studies, resource dependent groups could advocate for inclusion of traditional practices in governance frameworks. Ecological restoration including mangrove reforestation and coral reef rehabilitation remains a key strategy to buffer pollution impacts while enhancing biodiversity and coastal resilience. Concurrently, socioeconomic research must evaluate how pollution disproportionately affects marginalized communities, informing policies that balance environmental justice with economic development. Policy innovation should focus on source reduction through stricter regulations on industrial discharge and agricultural runoff, coupled with incentives for circular economies to minimize waste.

International cooperation is essential to address transboundary pollution particularly in shared fisheries and migratory species habitats. Public engagement campaigns, aligned with the UN Ocean Decade's goals, can bridge awareness gaps and foster community-led conservation initiatives. Prioritizing these interconnected strategies—grounded in equity, technological advancement, and ecosystem-based management—will be vital to achieving Sustainable Development Goal 14 and safeguarding coastal livelihoods in a rapidly changing climate.

Concluding comments

Coastal water pollution has emerged as a severe threat to both marine ecosystems and human health with plastics, fertilizers, chemicals, and oil spills significantly disrupting the balance of ocean biodiversity. This pollution not only endangers marine life through entanglement and ingestion but also introduces harmful chemicals like persistent organic pollutants (POPs) and heavy metals into the food chain posing risks to vulnerable populations including children and pregnant women. Given the vital role oceans play in sustaining human life, economic security, and global biodiversity, immediate and comprehensive action is essential. Tackling ocean pollution requires a multifaceted approach and technological innovations such as improved recycling, biodegradable materials, and advanced waste treatment systems. International cooperation, strict regulations, and increased public awareness are equally critical to reducing pollutant inflows. Community participation in conservation efforts is vital for creating a shared sense of responsibility to protect our oceans.

To achieve long-term sustainability, it is crucial that we intensify efforts to manage ocean ecosystems holistically.

Through coordinated global action, we can ensure that oceans continue to thrive for future generations, safeguarding both marine and human health. The future of a healthy marine environment depends on the choices we make today to reduce pollution and foster responsible stewardship.

References

1. Liquete C, Piroddi C, Drakou EG, Gurney L, Katsanevakis S, Charef A, et al. Current status and future prospects for the assessment of marine and coastal ecosystem services: A systematic review. Bograd SJ, editor. PLoS ONE. 2013 Jul 3;8(7):e67737.

2. Leal M & Spalding MD, editors. The state of the world's mangroves 2024. 2024 Jul. https://www.mangroveall iance.org/wp-content/uploads/2024/07/SOWM-2024-HR.pdf

3. Jean-Pierre Gattuso & Hansson L. Ocean acidification. Oxford: Oxford University Press. 2020.

4. Asthana S & Gibson A. Averting a public health crisis in England's coastal communities: a call for public health research and policy. Journal of Public Health. 2021 May 13;44(3).

5. Lisa Levin, *et al.* Global observing needs in the deep ocean. Front. Mar. Sci..2019. 10.3389/fmars.2019.00241

6. Rob Rowan, Nancy Knowlton, Andrew Baker & Javier

Jara, "Landscape ecology of algal symbionts creates variation in episodes of coral bleaching". Nature. 1997; 388, 265–269. doi: 10.1038/40843

7. Maodian Liu et al. "Observation-based mercury export from rivers to coastal oceans in East Asia". Environmental Science & Technology. 2021, 55 (20),14269 14280. https://doi.org/10.1021/acs.est.1c03755

8. Sheena Asthana & Alex Gibson. Averting a public health crisis in England's coastal communities: A call for public health research and policy. J Public Health (Oxf). 2021 May 13. https://pmc.ncbi.nlm.nih.gov/articles/PMC9424058/.

9. Ganachaud A & Wunsch C. Oceanic nutrient and oxygen transports and bounds on export production during the World Ocean Circulation Experiment. Global Biogeochemical Cycles. 2002 Oct 16;16(4):5–15–14.

10. Landrigan PJ. Human health and ocean pollution. Annals of Global Health. 2020 Dec 3;86(1):151. https://www.ncbi.nlm.nih.gov/pmc/articles/PMC7731724/

11. Madhav S, Ahamad A, Singh AK, Kushawaha J, Chauhan JS, Sharma S, et al. Water pollutants: Sources and impact on the environment and human health. Sensors in water pollutants monitoring: Role of material. Springer. 2019 Oct 12;43–62. https://link.springer.com/chapter/10.1007%2F978-981-15-0671-0_4.

12. US EPA. Arsenic rule small entity compliance guide for public water systems. US EPA. 2015. https://www.epa.g ov/dwreginfo/arsenic-rule-small-entity-compliance- guide-public-water-systems

13. Bryan BA & Kandulu JM. Designing a policy mix and sequence for mitigating agricultural non-point source pollution in a water supply catchment. Water Resources Management. 2010 Nov 25;25(3):875–92.

14. Central Ground Water Board. Report on status of ground water quality in coastal aquifers of India. Government of India, Ministry of Water Resources Central Ground Water Board Faridabad. 2014. https://www.cgwb.gov.in/ old_website/WQ/Costal%20Report.pdf

15. Prusty P & Farooq SH. Seawater intrusion in the coastal aquifers of India - A review. HydroResearch. 2020;3:61–74.

16. Ganoulis JG. Pollutant dispersion in oceans. Disorder and Mixing. 1988;139–42.

17. Metilelu OO, Adeniyi MO & Ekum MI. Modelling the dynamic effect of environmental pollution on coastal tourism. Scientific African. 2022 Sep 1;17:e01364–4.

18. US Environmental Protection Agency. Report to congress: Impacts and control of combined sewer overflows and sanitary sewer overflows. US Government Publishing

Office. 2004 Aug. https://www.epa.gov/npdes/2004-npdes-cso-report-congress

19. Central Pollution Control Board. COMAPS. CPCB. https://cpcb.nic.in/comaps

20. Agarwal S, Soumendra Darbar & Saha S. Challenges in management of domestic wastewater for sustainable development. Current directions in water scarcity research. Science Direct. 2022 Jan 1;531–52.

21. Lakshmanna B, Jayaraju N, Sreenivasulu G, Prasad TL, Nagalakshmi K, Kumar MP, et al. Evaluation of heavy metal pollution from coastal water of Nizampatnam Bay and Lankevanidibba, East Coast of India. Journal of Sea Research. 2022 Jun;102232.

22. Bhardwaj A, Kumar S & Singh D. Tannery effluent treatment and its environmental impact: A review of current practices and emerging technologies. Water Quality Research Journal. 2023 May;58(2):128–52. https://iwaponline.com/wqrj/article/58/2/128/94991

23. Central Pollution Control Board. Industry specific standards – Effluents/emissions. CPCB. https://cpc b.nic.in/effluent-emission/

24. Jonathan MP, S. Srinivasalu, N. Thangadurai, T. Ayyamperumal, Armstrong-Altrin JS & V. Ram-Mohan. Contamination of Uppanar River and coastal waters off Cuddalore, Southeast coast of India. Environmental

Geology. 2007 Apr 25;53(7):1391–404.

25. Verlecar XN, Desai SR, Sarkar A & Dalal SG. Biological indicators in relation to coastal pollution along Karnataka coast, India. Water Research. 2006 Oct;40(17):3304–12.

26. Berdalet E, Fleming LE, Gowen R, Davidson K, Hess P, Backer LC, et al. Marine harmful algal blooms, human health and wellbeing: Challenges and opportunities in the 21st century. Journal of the Marine Biological Association of the United Kingdom. 2015 Nov 20;96(1):61–91. https://www.ncbi.nlm.nih.gov/pmc/articles/PMC46762 75/

27. Food and Agriculture Organization of the United Nations. Water pollution from and to agriculture. International Decade for Action. 2017. https://wateractiondecade.org/ 2017/12/09/water-pollution-from-and-to-agriculture/

28. Yadav IC, Devi NL, Syed JH, Cheng Z, Li J, Zhang G, et al. Current status of persistent organic pesticides residues in air, water, and soil, and their possible effect on neighboring countries: A comprehensive review of India. Science of The Total Environment. 2015 Apr 1;511:123–37. https://www.sciencedirect.com/science/ article/pii/S0048969714017434

29. Syafrudin M, Kristanti RA, Yuniarto A, Hadibarata T, Rhee J, Al-onazi WA, et al. Pesticides in drinking water—A review. International Journal of Environmental Research

and Public Health. 2021 Jan 8;18(2). https://www.ncbi.nl m.nih.gov/pmc/articles/PMC7826868/

30. Kumar R & Dev K. Effects of chemical fertilizers on human health and environment: A review. International Advanced Research Journal in Science, Engineering and Technology ISO. 2007;3297(6). https://iarjset.com/ upload/2017/june-17/IARJSET%2036.pdf

31. Gwimbi P, George M & Ramphalile M. Bacterial contamination of drinking water sources in rural villages of Mohale Basin, Lesotho: exposures through neighbourhood sanitation and hygiene practices. Environmental Health and Preventive Medicine. 2019 May 15;24(1). https://www.ncbi.nlm.nih.gov/pmc/articl es/PMC6521341/

32. Manyi-Loh C, Mamphweli S, Meyer E & Okoh A. Antibiotic Use in agriculture and its consequential resistance in environmental sources: Potential public health implications. Molecules. 2018 Mar 30;23(4):795.

33. Ayyappan S & Jena J. Environmental issues in Indian freshwater aquaculture. Central Institute of Freshwater Aquaculture. https://eprints.cmfri.org.in/8579/1/Ayyap pan_Aquaculture_and_the_Environment.pdf

34. UNESCO. New UNESCO report: Rate of ocean warming doubled in 20 years, rate of sea level rise doubled in 30 years. UNESCO 2024. https://www.unesco.org/en/article s/new-unesco-report-rate-ocean-warming-doubled-2

0-years-rate-sea-level-rise-doubled-30-years

35. Doe J. Marine heatwaves and coral bleaching events. Journal of Climate Science. 2023;32(4):215–30.

36. Smith J. Impacts of climate change on coastal zones: Sea level rise and its consequences. Journal of Climate Science. 32(4):215–30.

37. Brown E. Phenological changes in marine ecosystems. Oceanography Review. 18(2):85–102.

38. Yuewen D & Adzigbli L. Assessing the impact of oil spills on marine organisms. Journal of Oceanography and Marine Research. 2018;06(01).

39. Qi D, Chen L, Chen B, Gao Z, Zhong W, Feely RA, et al. Increase in acidifying water in the western Arctic Ocean. Nature Climate Change. 2017 Mar 1;7(3):195–9. https://www.nature.com/articles/nclimate3228

40. McKee AM & Cruz MA. Microbial and viral indicators of pathogens and human health risks from recreational exposure to waters impaired by fecal contamination. Journal of Sustainable Water in the Built Environment. 2021 May;7(2):03121001

41. Omasete J, Forster J, Geere J, Drik W & Haque H. Water, sanitation and hygiene: The foundation for building resilience in climate-vulnerable communities. WaterAid. 5-12. http://www.indiaenvironmentportal.org.in/files/

file/water%20sanitation%20and%20hygiene.pdf

42. Syafrudin M, Kristanti RA, Yuniarto A, Hadibarata T, Rhee J, Al-onazi WA, et al. Pesticides in drinking water—A review. International Journal of Environmental Research and Public Health. 2021 Jan 8;18(2). https://www.ncbi.nl m.nih.gov/pmc/articles/PMC7826868/

43. Mitra S, Chakraborty AJ, Tareq AM, Emran TB, Nainu F, Khusro A, et al. Impact of heavy metals on the environment and human health: novel therapeutic insights to counter the toxicity. Journal of King Saud University, Science. 2022;34(3):101865. https://www.sc iencedirect.com/science/article/pii/S101836472200046 5

44. Mohammed Sadiq I, Chandrasekaran N & Mukherjee A. Studies on effect of TiO2 nanoparticles on growth and membrane permeability of Escherichia coli, pseudomonas aeruginosa, and bacillus subtilis. Current Nanoscience. 2010 Aug 1;6(4):381–7.

45. Shayo GM, Elimbinzi E, Shao GN & Fabian C. Severity of waterborne diseases in developing countries and the effectiveness of ceramic filters for improving water quality. Bulletin of the National Research Centre. 2023 Jul 24;47(1):113. https://rdcu.be/dlpCP

46. Semenza JC, Trinanes J, Lohr W, Sudre B, Löfdahl M, Martinez-Urtaza J, et al. Environmental suitability of vibrio infections in a warming climate: An early warning

system. Environmental Health Perspectives. 2017 Oct 3;125(10):107004.

47. Michel J & Fingas M. Oil spills: Causes, consequences, prevention, and countermeasures. World Scientific Series in Current Energy Issues. 2016 May 5;1:159–201.

48. Adelodun B, Ajibade FO, Ighalo JO, Odey G, Ibrahim RG, Kareem KY, et al. Assessment of socioeconomic inequality based on virus-contaminated water usage in developing countries: A review. Environmental Research. 2021 Jan 1;192:110309. https://www.sciencedirect.com/science/article/abs/pii/S0013935120312068?via%3Dihu b

49. US EPA. Drinking water regulations. US EPA. 2015. https://www.epa.gov/dwreginfo/drinking-water-reg ulations

50. Jal Jeevan Mission. Drinking water quality monitoring & surveillance framework Government of India Ministry of Jal Shak Department of Drinking Water & Sanita on Na onal Jal Jeevan Mission. Jal Jeevan Mission. 2021 Oct. https://jaljeevanmission.gov.in/sites/default/files/guideline/WQMS-Framework.pdf

51. E R I E S O G W S, Ta R. Prepared for: Technology Overview Report Phytoremediation. Ground-Water Remediation Technologies Analysis Center. 1996. https://www.clu-in.org/download/toolkit/phyto_0.pdf

52. Niti Aayog. Urban wastewater scenario in India. Niti Aayog. 2022 Aug. https://www.niti.gov.in/sites/default/files/2022-09/Waste-Water-A4_20092022.pdf

53. Afreen Nishat, Yusuf M, Qadir A, Yassine Ezaier, Vambol V, M. Ijaz Khan, et al. Wastewater treatment: A short assessment on available techniques. Alexandria Engineering Journal. 2023 Aug 1;76:505–16.

54. Oteng-Peprah M, Acheampong MA, deVries NK. Greywater characteristics, treatment systems, reuse strategies and user perception—A review. Water, Air, & Soil Pollution. 2018 Jul 16;229(8). https://link.springer.com/article/10.1007/s11270-018-3909-8

55. E R I E S O G W S, Ta R. Prepared for: Technology Overview Report Phytoremediation. Ground-Water Remediation Technologies Analysis Center. 1996. https://www.clu-in.org/download/toolkit/phyto_o.pdf

56. Kristanti RA & Hadibarata T. Phytoremediation of con-taminated water using aquatic plants, its mechanism and enhancement. Current Opinion in Environmental Science & Health. 2023 Apr;32:100451.

6

Understanding the Effectiveness of Action Plans on Air Pollution as a Critical Component of Health Policies in India

Abstract

Air pollution is a significant public health crisis. It is also considered the single largest environmental health threat with India experiencing some of the highest pollution levels in the world. Air pollution contributes to respiratory diseases and cardiovascular conditions which result in premature mortality. This review evaluates the effectiveness of action plans in addressing air pollution. Despite the formulation of regulatory frameworks such as the National Clean Air Programme (NCAP) and the National Ambient Air Quality Standards (NAAQS), air pollution continues to contribute to significant morbidity and mortality. This review highlights the gaps and challenges in implementation, including insufficient air quality monitoring, poor fund utilization, lack of transparency, and the absence of a robust governance mechanism. It also underscores that limited integration of air pollution mitigation strategies into public health policies, despite its direct impact on respiratory, cardiovascular, and other chronic diseases. Addressing these challenges requires a multi-sectoral approach that incorporates stringent policy enforcement, increased public awareness, improved data-driven decision-making, and enhanced inter-agency collaboration. Strengthening climate resilient health systems is imperative for mitigating the long-term health and socioeconomic consequences of air pollution in India.

Keywords: National Clean Air Programme, National Ambient Air Quality Standards, National Programme on Climate Change and Human Health, air pollution, action plan, and health impacts.

Authors

Drishya Pathak, Project Management Consultant for Central Drug Standard Control Organization with the Bill & Melinda Gates Foundation

Background

Since the 18th century, industrialization and human activities have significantly increased the production and emission of various air pollutants including nitrogen oxides (NO_x), sulfur dioxide (SO_2), carbon monoxide (CO), black carbon, ammonia (NH_3), and non-methane volatile organic compounds (NMVOCs) (1). Between 1990 and 2010, global annual emissions from fossil fuel combustion increased by approximately 50 percent, rising from six billion to nearly nine billion tonnes. Air pollutants are now major contributors to air pollution. Among the most concerning pollutants are nitrogen compounds, sulfur compounds, ozone, and particulate matter, which have far-reaching environmental and health consequences(2).

Global emissions have escalated dramatically with nitrogen oxide emissions reaching approximately 120 metric tonnes and carbon monoxide emissions exceeding 600 metric tonnes (1). Additionally, the intercontinental transport of particulate matter and ozone has become an emerging concern, exacerbating air quality issues beyond national boundaries. Despite initial underestimation of pollution growth, humanity now faces prolonged exposure to deteriorating air quality which poses severe threats to both environmental and public health.

Currently, an estimated 99 percent of the global population is exposed to air quality that exceeds the World Health Organization (WHO) guidelines (PM 2.5 concentration annual average <5 µg/m3) posing severe health risks. Despite efforts to monitor air quality in over 6,000 cities across 117 countries, residents continue to be exposed to hazardous levels of fine particulate matter (PM2.5) and nitrogen dioxide (3). This risk is increasing as there is a continuous increase in the number of polluted cities across the globe. Air pollution is now recognized as one of the most urgent environmental health crisis, contributing to approximately seven million premature deaths annually. An estimated 90 percent of the global population breathes polluted air, increasing the risk of respiratory diseases such as asthma, as well as cardiovascular diseases and lung cancer especially among populations in low- and middle-income countries that are exposed to high levels of air pollution (4).

There is substantial evidence highlighting the severity of air pollution and its far-reaching impacts on both climate change and public health—contributing to rising morbidity and mortality. There is clearly and urgent need to implement effective interventions. Numerous action plans have been implemented or are currently in development, with specific targets and timelines set for the coming decade. However, the success of these mitigation strategies depends on robust data analysis, which remains a significant challenge in addressing the problem of air pollution effectively.

In addressing the problem of air pollution, the availability and reliability of air quality monitoring is essential. While some countries have well established air quality surveillance systems, others are as yet in the early stages of installing monitoring infrastructure. This limitation hinders the accurate assessment of air pollution trends and the development of evidence-based policies to mitigate its impact. Some international platforms provide monthly and annual indices of air quality (AQ) for several countries. Organizations such as IQAir, the Climate and Clean Air Coalition (CCAC), the World Health Organization (WHO) ambient air quality database (since 2011), and World Data are among those platforms. However, prevention of air pollution needs air quality monitoring on a large scale. According to the latest data, the ten countries with the highest levels of air pollution worldwide—measured by PM2.5 concentrations exceeding hazardous thresholds—are in the Asia Pacific region. The primary pollutants in PM2.5 include black carbon and tropospheric ozone. Over 2.3 billion people are exposed to air pollution levels significantly above the WHO-recommended safety limits (5).

Notably, 83 of the 100 most polluted cities in the Asia Pacific region are in India—a statistic that has remained unchanged over the past five years. This persistent trend underscores the urgent need for effective mitigation measures and policy interventions to improve air quality nationwide. In this chapter, the author examines current and future strategies for air quality management in India where air pollution remains a critical challenge. It is, therefore, necessary to examine health concerns influencing regulatory frameworks and adaptation efforts within the national action plan on climate change and health.

Air quality standards

An air quality standard defines the maximum amount of a pollutant averaged over a specified period of time that can be present in outdoor air without harming public health. It thus defines clean air (6). These standards are based on the effect of air pollution on human health. Hence, regular air quality management is the only way to protect public health in the long-term for which standard monitoring tools are needed.

Air quality management strategies reviewed in the chapter are based on an existing regulatory framework called the National Ambient Air Quality Standards (NAAQS). The first NAAQS were introduced in the United States in 1971 (6). NAAQS establish limits on atmospheric concentrations of six major pollutants that contribute to smog formation, acid rain, and adverse health effects. These standards were developed by the U.S. Environmental Protection Agency (EPA) under the authority of the Clean Air Act and apply to outdoor air quality nationwide. The Central Pollution Control Board (CPCB) of India adopted NAAQS in the 1980s and delineated the following objectives;

- To indicate the levels of air quality necessary with an adequate margin of safety to protect public health, vegetation, and property.
- To assist in establishing priorities for the abatement and control of pollutant levels.
- To provide a uniform yardstick for assessing air quality at the national level.
- To indicate the need and extent of the monitoring program (7).

The NAAQS are designed to safeguard both public health and the environment, with two distinct categories of standards: primary standards, which protect vulnerable populations such as children, the elderly, and individuals with respiratory conditions, and secondary standards which focus on preventing environmental and property damage (8).

Six criteria air pollutants (CAPs) refer to (Table 1) regulated under the NAAQS include ozone (O_3), particulate matter ($PM_{2.5}$ and PM_{10}), lead (Pb), carbon monoxide (CO), sulfur oxides

(SO_x), and nitrogen oxides (NO_x) (9). Sources of PM in the atmosphere include stationary and mobile emissions mainly due to the combustion of fossil fuels and mechanical grinding or crushing activities. There are other sources as well including sand, biogenic emissions, and sea spray. The fine particles stay in the atmosphere and are transported from one region to another, whereas PM_{10} particles settle naturally by mechanisms involving gravitation or wet scavenging (10).

This chapter specifically examines the regulation and management of particulate matter ($PM_{2.5}$ and PM_{10}), given its significant impact on human health and air quality.

Table 1	Six criteria of air pollutants (CAPs) regulated under NAAQS and associated human health risks				
Pollutant	Primary/ Secondary	Averaging Time	Level	Form	Human Health Risks
Carbon monoxide (CO)	Primary	8 hours	9 ppm	Not to be exceeded more than once per year	Exacerbates symptoms of heart disease such as chest pain; may cause vision problems and reduce physical and mental capabilities in healthy people
		1 hour	35 ppm		
Lead (Pb)	Primary and secondary	Rolling three month average	0.15 µg/m3 (1) (2)	Not to be exceeded	Adverse effects on multiple bodily systems; may contribute to learning disabilities when young children are exposed; cardiovascular effects in adults
Nitrogen dioxide (NO₂)	Primary	1 hour	100 ppb	98th percentile of 1-hour daily maximum concentrations averaged over 3 years	Inflammation and irritation of breathing interferes with the ability of certain plants to respire leading to increased susceptibility to other environmental stressors (e.g., disease, harsh weather)
	Primary and secondary	1 year	53 ppb (9) (2)	Annual mean	

Ozone (O_3),		Primary and secondary	8 hours	0.070 ppm (9)(3)	Annual fourth-highest daily maximum 8-hour concentration averaged over 3 years	Reduced lung function; irritation and inflammation of breathing passages
Particle pollution (PM)	PM 2.5	Primary	1 year	9.0 µg/m3	Annual mean averaged over 3 years	Irritation of breathing passages, aggravation of asthma, irregular heartbeat
	PM 2.5	Secondary	1 year	15.0 µg/m3	annual mean averaged over 3 years	
	PM 2.5	Primary and secondary	24 hours	35 µg/m3	98th percentile averaged over 3 years	
	PM 10	Primary and secondary	24 hours	150 µg/m3	Not to be exceeded more than once per year on average over 3 years	
Sulfur dioxide (SO2)		Primary	1 hour	75 ppb (9)	99th percentile of 1-hour daily maximum concentrations averaged over 3 years	Breathing difficulties, particularly for people with asthma and heart disease
		Secondary	1 year	10 ppb	Annual mean averaged over 3 years	

Source: U.S. Environmental Protection Agency

Pollution levels in India

According to the Global Health Observatory Data Repository, the average $PM_{2.5}$ exposure in India is estimated at 65.2 µg/m³, significantly exceeding the National Ambient Air Quality Standards (NAAQS). Air pollution risks are typically quantified for ambient particulate matter pollution, household air pollution, and to a smaller extent, tropospheric ozone. In India, the main sources of particulate matter pollution are residential and commercial biomass burning, windblown mineral dust, coal burning for energy generation, industrial emissions, agricultural stubble burning, waste burning, construction activities, brick kilns, transport vehicles, and diesel generators (11). A study was initiated by Source Apportionment Studies in 2007 across six Indian cities (i) Delhi, (ii) Mumbai, (iii) Chennai, (iv) Bangalore, (v) Pune, and (vi) Kanpur with the objective to profile ground level concentration (GLC) of air pollutants in different parts of the city including background, residential, commercial/mixed areas and source specific "hot spots" viz. kerbside/roadside, industrial zones, etc. The objective was to arrive at appropriate emission factors (EF) for different categories of vehicles and non-vehicular sources, prepare an inventory of different air pollutants, profile sources of emission, and assess the impact of sources on ambient air quality across these cities (12). The report showed that over the years PM_{10} and $PM_{2.5}$ levels in ambient air remained consistently high regardless of location type. Data indicators from 2007 showed that Chennai and Bangalore had better air quality for all PM parameters. The study also suggested that for residential areas, air quality standards are exceeded by over 90 percent of the time with respect to 24-hourly average standard for PM_{10} except for Chennai and Bangalore (12). For $PM_{2.5}$, standards are exceeded 100 percent of the time at kerb stations and industrial and residential areas of all except for these two cities. However, a lot has changed since this data was published a decade ago. The following data indicators were published in the report;

· Delhi showed high variability in PM levels

- NO_2 levels exceeded in residential area sites of Delhi (35%), Pune (6%), and Mumbai (25%).
- The winter and post-monsoon seasons were the most critical periods with pollution levels exceeding standards at a much higher rate compared to the summer months.
- In addition to particulate matter, nitrogen dioxide (NO_2) was an emerging pollutant of concern (13)(14).

The global database also highlighted similar patterns. The WHO Air Quality Database (2018) reported that nine of the ten most polluted cities in the world are in India. These are Kanpur, Faridabad, Gaya, Varanasi, Patna, Delhi, Lucknow, Agra, and Gurgaon (14). These data points highlight the urgent need to link public health to air pollution and to take effective action.

Air pollution: A threat to human health

Besides being the largest environmental health threat globally, air pollution is also one of the major causes of premature death. Extensive research has established a clear link between air pollution and its adverse health effects. Several studies have shown that short-term and long-term exposure to air pollution is associated with mortality and morbidity. The Lancet Commission on Pollution and Health reports air pollution as a critical risk factor for noncommunicable diseases causing an estimated one-quarter of all adult deaths from heart diseases, 25 percent from stroke, 43 percent from chronic obstructive pulmonary disease, and 29 percent from lung cancer (15)(14).This data is drawn from the total of 6.5 million deaths reported annually, attributed to various sources of air pollution. Air pollution is found to be associated with lung disease, cardiovascular disease, and diabetes which is nearly 40 percent of the disease burden due to air pollution in India. $PM_{2.5}$ exposure has been related to low birthweight, chronic kidney disease, and neurodegenerative diseases according to global studies. These studies have not been conducted in India so far (16) .

In light of the international estimates, there is a high burden of disease due to air pollution, India felt it was necessary to take stock of data from Indian studies on the health impact of air pollution. A high-level committee was constituted in the Ministry of Health & Family Welfare (MoHFW) to develop a compendium of studies on the health impact of air pollution in the Indian context (14). This study captured data from research conducted by various stakeholders such as the Indian Council of Medical Research (ICMR), the National Institute of Occupational Health (NIOH), Central Pollution Control Board (CPCB), State Pollution Control Boards (SPCB), WHO, The Energy and Resources Institute (TERI), Post Graduate Medical Institute of Medical Education and Research (PGI), Chandi-garh, and the All India Institute of Medical Sciences (AIIMS), Delhi. Several action points emerged from the studies for the Ministry of Health and Family Welfare (MoHFW), including the establishment of a high-level task force to mitigate the effects of air pollution, strengthen advocacy efforts, enhance information, education, and communication (IEC) initiatives for public awareness, and integrate air pollution mitigation strategies with various national health programs (14).

In 2018, the ICMR conducted a comprehensive study, publishing the first detailed estimates of the impact of air pollution's on health loss and life expectancy reduction across Indian states. This report was released under the India State-Level Disease Burden Initiative, a collaborative effort involving ICMR, the Public Health Foundation of India (PHFI), and the Institute for Health Metrics and Evaluation (IHME). The initiative was carried out in partnership with the Ministry of Health and Family Welfare in over 100 Indian research institutions (17). The report stated:

- India has a disproportionately high percentage of global premature deaths and a high disease burden due to air pollution
- One in eight deaths in India were attributed to air pollution in 2017 i.e. 1.7 million deaths— some of which were due to chronic diseases including chronic obstructive pulmonary disease, ischaemic heart disease, stroke, and lower respiratory infections (18)
- 6.7 lakh deaths were due to outdoor particulate matter air pollution
- 4.8 lakh deaths were due to household air pollution
- In 2017, 77 percent of the population of India was exposed to ambient particulate matter of $PM_{2.5}$ and above. The mean annual exposure was 90 µg/m3 which is one of the highest in the world and is way above the recommended value of 40 µg/m3.
- The disability-adjusted life year (DALY) rates varied six-fold for outdoor particulate matter. Some of the chronic lung diseases were COPD (22.7%), lower respiratory infections (15.5%), and lung cancer (1.3%) (18)
- Average life expectancy in India would have been 1.7 years

higher of the air pollution level were less than the minimal level.

These alarming statistics on air pollution-related morbidity and mortality in India underscored the urgent need for designing comprehensive policy interventions. With millions of premature deaths and a significant disease burden linked to air pollution, the country is facing a critical public health challenge that requires coordinated efforts across multiple sectors. In response, the National Clean Air Programme was introduced as a major step to achieve one of the key Sustainable Development Goals for reduction of the burden of deaths and diseases from air pollution. It was a long journey for India to work on a specific action program that addresses the issue of rising air pollution (Figure 1).

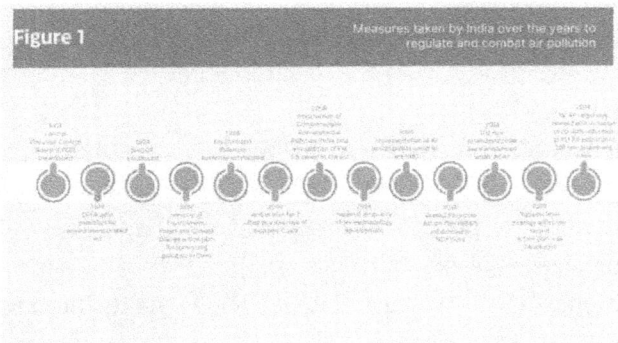

Figure 1 — Measures taken by India over the years to regulate and combat air pollution

Source: Compiled by authors

National Action Plan for Climate Change and Human Health

India introduced the National Action Plan on Climate Change

(NAPCC) on June 30, 2008, identifying eight core "national missions" to outline existing and future policies for addressing climate mitigation and adaptation. In 2014, following the Conference of Parties (COP 21), the plan was revised to expand its scope, introducing four new missions including one dedicated to health. The National Center for Disease Control (NCDC) was designated as the nodal agency for drafting the Action Plan under the Health Mission. Collaborating with key stakeholders such as Directorate General of Health Services (Dte.GHS), Ministry of Health and Family Welfare (MoHFW), Ministry of Environment, Forest & Climate Change (MoEFCC), Indian Council of Medical Research (ICMR), Department of Science and Technology (DST), National Disaster Management Authority (NDMA), Central Ground Water Board (CGWB), the Ministry of Agriculture, Central Pollution Control Board (CPCB), Ministry of Earth Sciences (MoES), The Energy and Resources Institute (TERI), and National Environmental Engineering Research Institute (NEERI), NCDC developed the National Action Plan on Climate Change and Human Health (NAPCCHH) (19).

Recognizing India's diverse geography and climate, the need for state-specific action plans was emphasized. The NAPC-CHH was designed to strengthen the health resilience of Indian citizens particularly of vulnerable populations including children, women, and marginalized communities against climate illnesses. The plan includes a range of strategic actions such as raising awareness among the general public, healthcare providers, and policy-makers; enhancing the capacity of the healthcare system through training, workshops, and curriculum development; improving health preparedness and response through situational analysis at national, state, and district levels; fostering partnerships and aligning efforts with other missions; and strengthening research capacity. This plan was approved by the National Health Mission Empowered Programme Committee (EPC) in November 2018 and National Health Mission Steering Group in February 2019 to support states to implement the NPCCHH.The initial desired inputs were to establish an Environmental Health Cell within the State Health Department and to employee State Nodal Officers (SNOs) for climate change within the State Health Department. The first budget allotted under the NPCCHH program for the FY 2021-22 was Rs. 7.5 crores and for next six year (2021 to 2026), it was Rs. 198 crores.

The NPCCHH is designed to addresses all health-related problems caused by climate change. However, this chapter specifically focuses on the objectives and outcomes related to air pollution. The first report highlighted key outcomes under each objective (20):

1. Creating awareness:

- Regular advisories were issued to all states on acute respiratory illnesses (ARIs).
- Special observances, such as for the World Environment Day and the first International Day of Clean Air for Blue Skies, were marked to raise awareness.
- The development of IEC materials and social media campaigns on platforms like Twitter.
- Public access to the Air Quality Index (AQI) for all cities was made available on the NCDC website.

2. Strengthening healthcare capacity:

- Training of 230 hospital nodal officers from sentinel hospitals across 24 states and four union territories on ARI surveillance related to air pollution.
- Training of state and district nodal officers for climate change from Madhya Pradesh, Kerala, Mizoram, Rajasthan, and Haryana.
- Development of community-level training modules for women and children.
- Inclusion of "Climate Resilient" and "Green" principles in healthcare facilities approved under Indian Public Health Standards (IPHS) standards within the National Health Mission.

3. Enhancing health preparedness and response:

- Surveillance of ARI in six central hospitals in Delhi and 317 hospitals across 23 states.
- Vulnerability and needs assessment conducted in three states.
- A national review meeting of all state nodal officers was held online to evaluate situational analysis at national, state, and district levels.

4. Developing partnerships and collaboration:

- Formation of a high-level inter-ministerial committee chaired by the Secretary of Health and Environment.
- Collaboration with national health programs to integrate climate and health-specific inputs and create synergy

with other missions.

5. Strengthening research capacity:

- Support extended to 30 research institutions including ICMR for air pollution studies (20).

To run parallel to the NAPCCHH, the National Clean Air Programme (NCAP) was launched as a monitoring plan for air pollution mitigation (21). This program designated 131 cities the status of non-attainment based on their capacity to meet the NAAQS for fine particulate matter (PM_{10}) that has a diameter of 10 microns or less or nitrogen dioxide over a five-year period (22).

National Clean Air Programme

India's flagship program for better air quality in 132 cities, launched in 2019, called the National Clean Air Programme (NCAP) is a collection of regulations, policies, and programs which aims to improve air quality and public health by identifying cost-effective measures to reduce emissions from all the known sources (23). The objective of the NCAP is to reduce $PM_{2.5}$ and PM_{10} levels by 20-30 percent by 2024 (later extended to 2026) aiming to have a 40 percent reduction as compared to 2017 in 132 cities in 23 states and two union territories (24).

The NCAP has three main segments that include mitigation actions, knowledge and database augmentation, and institutional strengthening (24). It has defined the following activities;

(a) Preparation of an information baseline for emissions and pollution loads and an assessment of source contributions in non-attainment cities.

(b) An air information cell to maintain and disseminate information generated under NCAP.

(c) Establishment of a technology assessment cell to support bilateral and multilateral agreements undertaken by the boards.

(d) Development of the network of technical institutions to provide support for policies and program of the Government of India on air pollution.

(e) Constitution of a three-tier mechanism at the pollution control boards to review assessment and inspection guidelines for implementation of standards.

(f) Implementation of an awareness, training, and capacity-building drive at the boards.

(g) Constitution of a committee to review the ambient and emission standards.

(h) Development of a framework to establish international cooperation to share best practices on air pollution (23).

NCAP was launched with the goal of improving air quality

in India by implementing city-specific and state-level interventions. A review of the monitoring indicators from CREA's annual reports (2022–2024) out of the seven thematic areas that the NCAP covers including air quality data, public outreach, capacity building, monitoring networks, and source apportionment highlights key areas like implementation, transparency, and effectiveness of various measures; (table 2).

Table 2	List of indicators under NCAP concerning air quality monitoring and their status			
Monitoring Indicators	2022 report by CREA	2023 report by CREA	2024 report by CREA	Data Visibility on PRANA
Funding				
Funds released	Rs.4,400 crores funds released and Rs. 1,157 crores utilized of 49 million plus cities plan	Rs. 652.62 crores released since 2019, and Rs 6,436 crores allocated under XVFC for the financial year 2020-2021 and 2021-2022	Rs.498 crores utilized of the Rs.1,253 crore allocated. Also, 5,909 crores utilized out of Rs. 9,610 crores allocated under 49 million plus cities plan	Yes, visible on the PRANA portal but only as a cumulative figure
Action Plan				
Cities with action plans	Yes, all 131	Yes, all 131	Yes, 131 but most of them have the same plans	Yes, visible on the PRANA portal
Emergency Response System (GRAP)	19 states and union territories where GRAPS is fully prepared. In other states, the Emergency response system (ERS) is still ongoing in various phases	No information	GRAP is prepared for all the non-attainment cities but it is state-specific	Yes, visible on the PRANA portal

Station action plans	No information	No information	No information	No information on PRANA
Regional action plans	No information	No information	Commission for air quality management (CAQM), responsible for regional air quality for Delhi-NCR exists. No institution or commission similar to CAQM exists for any other city	No information
Transboundary action plans	No information	No information	No information	No information
Studies under NCAP				
Source apportionment studies	15 cities in 9 states have completed studies	37 of 131 have completed source apportionment studies	44 of 131 have conducted source apportionment studies	Yes, visible on the PRANA portal
Capacity study	Not completed	Not completed	None of the cities completed carrying capacity studies	31 cities make no mention of carrying capacity study

National health profile and database	No information	No information	No information	No
Air pollution impact on health and economy	3 studies (two awarded, one completed	No information	No information	No
Activities				
Augment for PM 10 manual monitoring stations to 1,500 from 703	818	883 in 378 cities	As of Dec 2023, only 931 manual stations were operational	Yes, a list is present on the PRANA portal
2-3 average number of continuous ambient air quality monitoring system CAAQMS (per city)	309	401	531 CAAQMS operational across 279 cities, way above the target of 150.	Not on PRANA but visible on the CPCB website
Augment $PM_{2.5}$, monitoring station for all cities under NAMO	262 stations in 121 cities	NA	360 stations	Hyperlinked to CPCB portal

Plan for setting up 10 city-super networks	Under process	NA	No hotspot-based city forecasting system deployed	NA
Plan for setting up an air information center	Under process	NA	Only Delhi has a functional air pollution forecasting system called the central control room for air quality monitoring-Delhi-NCR (CCR: Delhi-NCR region)	Only the central air information system available called 'PRANA' portal
Online continuous emission and effluent monitoring system (OCEMS)	Proposed in 2023	Proposed in 2023	Promoting real-time monitoring of industrial emissions and effluent in India directly to CPCB servers and SPCBs. Only 17 state facilities collecting data out of 36 states and UTs.	

Source: Centre for Research on Energy and Clean Air Report 2022, 2023, and 2024 (25)(26)(27).

Points of discussion

Air pollution is one of the most important development challenges in India. While the NCAP is limited to 131 non-attainment cities, there is evidence that the problem extends far beyond these areas affecting its most crucial resources: its people, and its productivity (28). The population of India is being exposed to the highest levels of polluted air but is managed by weakly monitored action plans. The NCAP, launched to curb pollution in non-attainment cities, has made progress in terms of policy formulation and fund allocation. The CPCB has existed for decades with mandates for air pollution control under national and state legislation. However, a detailed review reveals critical gaps in its execution, transparency, and monitoring. Although, the NAPCCHH has been in operation for a decade and promises to address a wide range of health issues related to climate change and specifically air pollution, there is a lack of continuous efforts to implement those actions.

Policy and institutional gaps

All 131 non-attainment cities continue to have action plans. But there has been little revision or enhancement in their strategies over the years. The plans remain largely the same, raising concerns about their adaptability to evolving air pollution trends (25). Only eight states with 48 cities have specific action plans of their own. Also, there are no time bound reduction targets as part of these policy action plans.

There is increasing recognition of air pollution as a regional issue but there is a lack of concrete regional and transboundary action plans. The only notable initiative is the Commission for Air Quality Management (CAQM) which oversees air quality in Delhi-NCR. However, no equivalent institution exists in other regions leaving large parts of the country without coordinated regional interventions. Additionally, transboundary pollution—a critical issue in the border states—remains unaddressed with no reported action plans in CREA's assessments.

A major challenge in NCAP's implementation is the lack of comprehensive governance mechanisms at different levels—city, district, state, and regional (airshed level). Mitigation action plans have been developed by states but there is no interstate coordination which is due to the lack of an integrated framework that oversees air quality management holistically (26).

Currently, the approach to climate change as seen by the NAPCC and NAPCCHH is too broad and lacks specificity. To address these challenges, funding that supports multi-jurisdiction collaboration which will have higher leverage than funding that simply duplicates the existing infrastructure is needed. Also, mitigation and adaptation components in the form of quantified targets will yield better outcomes (29). In 2013, when China exceeded 52 µg/m3 concentration of $PM_{2.5}$, the Chinese government implemented comprehensive policies to reduce PM concentrations by 10 percent within five years to combat air pollution. When the population-weighted the annual average $PM_{2.5}$ concentration was calculated in 2018, there was a sharp 29 percent reduction compared to 2013 levels (30). This indicates that setting up quantified targets provides better outcomes.

Monitoring and transparency

One of NCAP's key targets is to expand air quality monitoring networks across cities. While some progress has been made, it remains below expectation. There are not enough stations to cover a territory or population of this size. The available number of stations are not even in line with CPCB's own guidelines (25). Expansion of $PM_{2.5}$ monitoring stations is necessary, especially in smaller cities, and more CAAQMS are needed that provide detailed city-wise data on the PRANA portal.

Despite these advancements, real-time monitoring of industrial emissions remains limited. Only 17 out of 36 states and union territories are collecting industrial emissions data under the Online Continuous Emission and Effluent Monitoring System (OCEMS) indicating that there is a critical gap in regulating industrial pollution sources.

Similarly, carrying capacity studies—which assess whether cities can sustain their current pollution levels—remain incomplete. In 2024, none of the cities had completed these assessments and 31 cities did not even mention any studies. The absence of research limits the ability to design scientifically backed pollution control measures and also limits the calculation of data on emission load reduction.

Additionally, data transparency remains a concern. While some information is available on PRANA, key datasets—such as station-specific action plans, regional air quality data, and year-wise fund utilization—are either missing or are presented in an aggregated form making detailed analysis difficult. One of the key objectives of the NAPCCHH is to focus on information, education, and communication (IEC) with social media as a primary activity. However, there is a noticeable gap in keeping information up to date. The Ministry of Health and Family Welfare, as part of the NPCCHH released a health advisory on air pollution on the NCDC website regarding poor air quality. This was last released on the website in 2022 (31). Social media accounts such as Twitter had posts dating back to June 3, 2023 with no recent updates (32). The lack of regular updates makes it difficult to make timely revisions based on ongoing air quality monitoring.

Emission inventories

Progress in source apportionment studies which identify pollution sources has been slow. With no access to these studies, it would be difficult to identify the source of air pollution and assess opportunities to improve air quality. By 2024, only 44 of the 131 non-attainment cities had completed the studies. There is, therefore, a lack of crucial data to guide mitigation strategies. Identifying the sources of air pollution and assessing opportunities to improve air quality is important. Emission factors and sources are either limited to a few cities or to a few pollutants. Most of the measurements are based on the studies conducted in North America and Europe (16).

Additionally, there is a significant gap in linking air pollution to health and to the economy. Only three studies were initiated in 2022. There is lack of sustained research efforts. India needs to speed up the process of aggregating and processing data to develop more accurate emission inventories with the help of public, private, and academic stakeholders.

Even under the most optimistic climate mitigation scenarios, carbon emissions and other pollutants will continue to accumulate in the atmosphere. Without substantial intervention, the environmental and health impacts of pollution are expected to intensify exacerbating serious health conditions. A study by Crooks L. James et al. (2021) on the ozone climate penalty, NAAQS attainment, and health equity along the Colorado Front Range found that residents were experiencing the effects of climate change including rising temperatures and increased ozone levels. This study underscored significant health equity concerns revealing that the burden of ozone pollution disproportionately affects historically disenfranchized and frontline communities in the non-attainment cities. Furthermore, it projected a delay of approximately two years in achieving NAAQS standards across the region, highlighting the urgent need for targeted policy interventions (33). This underscores the urgent need for implementing aggressive climate mitigation and adaptation strategies that prioritize vulnerable communities. Addressing these disparities requires the development of state-specific, city-specific, and region-specific ambient air quality action plans tailored to local environmental conditions and public health needs.

The governance framework for air quality management remains fragmented, lacking a comprehensive mechanism across city, district, state, and regional levels. Additionally, within NCAP cities, 78 exceeded the PM10 levels above the NAAQS and 118 non-NCAP cities also surpassed the safety limits highlighting widespread non-compliance. Despite the urgent need for capacity-building and public outreach, only six percent of the total funds have been allocated toward capacity development. A mere one percent has been spent on public awareness efforts.

These systemic issues underscore the need for stronger accountability, better fund utilization, enhanced data transparency, and a robust institutional framework to make meaningful improvements in air quality management.

Adaptive policies and responses

A review of the literature on NCAP suggests that several Indian policies and action plans acknowledge the impact of climate change on health. A major concern is that these action plans lack a strong emphasis on implementation strategies aimed at developing climate resilient health systems. Also, due to India's diverse climatic zones, it remains highly vulnerable to a wide range of climate-induced health challenges including heat waves, vector-borne diseases, and air pollution. One of the key adaptive policy responses introduced by the NCAP and CPCB for the health sector is the Graded Response Action Plan (GRAP). This plan provides a structured framework for implementing air pollution control measures based on the severity of air quality levels, ensuring timely and coordinated action to mitigate health risks. According to NCAP all states have developed GRAP for each of their cities but the issue is that all cities do not have dedicated GRAP specific to the city and public health related to the quality of air.

Developing climate resilient health systems requires a comprehensive approach to strengthening healthcare services, particularly in addressing air pollution related health challenges. Key priorities include:

Public awareness and education: Enhancing outreach efforts through multilingual channels is essential to raise community awareness of air pollution related health risks while considering different income groups. Research from China highlights the link between objective air pollution levels and public awareness of environmental issues, as well as the relationship between health status and environmental consciousness. One study found that individuals experiencing declining health are more likely to perceive air pollution as a threat and take adaptive measures. Additionally, it suggested that experiencing positive health improvements—such as feeling healthier than the previous year— correlates with higher productivity and increased workforce participation (34).

Education and awareness at multiple levels is crucial as objectively measured air pollution does not automatically translate into individual perceptions of environmental risks. These perceptions are shaped by personal experiences, values, and societal goals. Therefore, mass public awareness campaigns should be implemented. Community centered initiatives and public lectures can help generate environmental consciousness in lower-income groups. For middle- and high-income segments, engagement through in-depth professional reports, seminars, and workshops can foster a deeper understanding of environmental challenges and a stronger commitment to conservation efforts. A study by Abhijith et al in 2024 suggests the importance of a citizen science approach that engages the public to co-design effective solutions for air pollution challenges (35). Additionally, educating these groups about the adverse health effects of pollution would empower them to recognize and mitigate potential health risks associated with environmental degradation.

Integrated data reporting and quantifying health impacts: Timely reporting of air pollution-related illnesses to state and central health authorities is essential for evidence-based policy-making and swift action (18). Existing reporting formats under NPCCHH for acute respiratory illness (ARI) surveillance include daily or monthly hospital reports and graphical analysis formats for districts. However, there is a lack of clarity on how effectively these formats are being designed and being used as monitoring tools (36).

A recent report published by Climate and Health Reliance emphasizes the importance of quantifying the health impacts of air pollution and providing evidence-based action using technology that can be termed as Health Impact Assessment (HIA). A case study from Trindad and Tobago that faced significant air quality issues used AirQ+, a specialized tool developed by WHO, to estimate the burden of disease attributed to air pollution integrating air quality data with health statistics. The use of this technology helped health professionals and policy-makers translate air pollution exposure to tangible health outcomes including mortality and morbidity associated with specific pollutants. This data helped in creating a roadmap with key lines of action including infrastructural and technological improvements, specifically targeted capacity building activities, and a robust plan for monitoring and evaluation (37).

Health sector surveillance and response mechanisms: While HIA identifies the health burden of air pollution, Health System Assessment (HAS) evaluates the capacity of healthcare facilities and other sub-heads like logistics/equipment, human resources, access to essential medicines and diagnostics, and operational practices within health facilities which can be helpful to convert these facilities to "Green Facilities". HAS can also help in professional training, targeted programs, and capacity building on subject development of IEC. In the State of Chhattisgarh HAS was performed by the State Health Resource Center in 2022. This case study highlighted the importance of evaluation of the region's health system and to understand its readiness to address air pollution related health harms at these facilities. Based on this assessment, targeted interventions were introduced to upgrade the infrastructure, develop the workforce, and develop an efficient logistics framework by integrating a stock-monitoring system including respiratory medicines in the Essential Drug List of health facilities to ensure the availability of essential medicines for respiratory diseases (37).

Capacity-building at district and city levels: Preparedness of the healthcare sector can be enhanced by training health professionals to manage climate and pollution-related illnesses effectively. This requires engaging State Nodal Officers (SNOs) at the state level and district level. Although the NAPCCHH aims to build capacities of SNOs, engagement of community level stakeholders including NGOs and the local community is equally important. A task force of District Nodal Officers (DNOs) should be organized to do this. The NCDC website provides information on workshops conducted for DNOs but the frequency of such workshops is extremely low (38). HAS also offers a solution to such issues as it can help in supporting the capacity of the communities to regularly conduct air quality and heat monitoring to generate citizen science data for policy interventions (37). Another case study of Accredited Social Health Activists (ASHAs) from India showed how educating community level health workers can be a win-win situation in creating awareness programs at district and block level (37). Communities are then able to organize and advocate for stricter enforcement of environmental regulations.

Sentinel hospital networks: There are no specific studies or surveys done in India to understand how well-informed health professionals are. However, surveys conducted in other parts of the world with high rates of air pollution clearly show that public health professionals do not counsel patients or engage in community outreach on the health effects of air pollution and are not competent to address these adverse health effects related to air pollution at the individual as well as population level (39). Therefore, in countries like India, at locations with high rates of air pollution, there is a need for establishing three to four sentinel hospitals in each city to manage air pollution-related cases and serve as focal points for surveillance and treatment.

A possible solution could be implementation of an early warning system by seeking partnerships with environmental protection agencies and local and international academic researchers and to disseminate real-time information to communities facing high levels of air pollution. A similar case study on the medical health community in the UK showed that their strategy could protect patients from the adverse effects of air pollution. A group of four medical institutions fostering a collaborative approach use a toolkit to integrate knowledge about air pollution into clinical practice to help healthcare professionals communicate effectively with patients about air quality. They also produced a combined report in 2018 highlighting the severe health impacts of air pollution across the lifetime. This group of institutions provided personalized care for high-risk patients by developing personalized care plans (37).

Concluding comments

This review shows that while several efforts have been made to tackle air pollution in India, major gaps persist in policy execution, enforcement, and inter-sectoral coordination. The National Clean Air Programme (NCAP) provides an essential foundation but its success remains limited due to inadequate fund utilization, the lack of comprehensive monitoring mechanisms, and the absence of time-bound reduction targets for many cities. Enhancement of capabilities is recognized as an essential priority within NCAP. Training programs should be designed to improve the skills of workers in charge of air quality management and to strengthen the responsibilities of pollution control boards. But due to the lack of stringent norms there has been no accountability of the specified targets. The lack of real-time air quality data, source apportionment studies, and regional coordination frameworks has weakened the ability to develop targeted interventions.

The public health dimension of air pollution remains under-addressed. Despite NAPCCHH, the national health program, covering air pollution as a major mission, the necessary goals required like data to support policy development, quantification of the health burden to drive accountability, and integration of parallel running programs to engage different stakeholders are missing. These are absolutely essential to enhancing health sector capacity.

To achieve meaningful progress, it is important to strengthen air pollution governance at the city, state, and national level and ensure greater transparency, accountability, and inter-sectoral collaboration. Additionally, capacity-building initiatives, public awareness campaigns, and real-time data monitoring must be prioritized to support evidence-based policy-making. Without decisive action, air pollution will continue to pose a significant threat to both environmental sustainability and public health in India.

References

1. Ritchie H & Roser M. Air pollution: Our overview of indoor and outdoor air pollution. Our World Data. 2021 Jan 1. https://ourworldindata.org/air-pollution

2. European Environment Agency. Increasing environmental pollution (GMT 10). European Environment Agency. 2015 Feb 18. https://www.eea.europa.eu/soer/2015/global/pollution

3. World Health Organization. Billions of people still breathe unhealthy air: New WHO data. World Health Organization. 2022 Apr 04.https://www.who.int/news/item/04-04-2022-billions-of-people-still-breathe-unhealthy-air-new-who-data

4. United Nations Environment Programme. Five cities tackling air pollution. UNEP. 2022 Sep 06. https://www.unep.org/news-and-stories/story/five-cities-tackling-air-pollution

5. Climate & Clean Air Coalition. Air pollution measures for Asia and the Pacific. Climate & Clean Air Coalition. https://www.ccacoalition.org/content/air-pollution-me asures-asia-and-pacific

6. California Air Resources Board. National ambient air quality standards. California Air Resources Board. https://ww2.arb.ca.gov/resources/national-ambient -air-quality-standards

7. California Air Resources Board. Southern California wildfire recovery. California Air Resources Board. 2019. NAAQS_2019.pdf. https://cpcb.nic.in/upload/NAAQS_2 019.pdf

8. National Ambient Air Quality Standards (NAAQS). Definition, criteria pollutants, & facts. Britannica. 2025. https://www.britannica.com/science/National-Ambient -Air-Quality-Standards-United-States

9. US EPA O. Criteria Air Pollutants: NAAQS Table. EPA. 2014: https://www.epa.gov/criteria-air-pollutants/naa qs-table

10. Gopikrishnan GS & Kuttippurath J. Impact of the National Clean Air Programme (NCAP) on the particulate matter pollution and associated reduction in human mortalities in Indian cities. Sci Total Environ. 2025 Mar 10;968:178787. https://www.sciencedirect.com/science/ article/pii/S004896972500422X

11. India State-Level Disease Burden Initiative Air Pollution Collaborators. Health and economic impact of air pollution in the states of India: the Global Burden of Disease Study 2019. Lancet Planet Health. 2020 Dec 27;5(1):e25–38. https://www.ncbi.nlm.nih.gov/pmc/arti cles/PMC7805008/

12. Central Pollution Control Board. Air quality monitoring, emission inventory and source apportionment study for Indian cities. Central Pollution Control Board. 2011 Feb. displaypdf.pdf. https://cpcb.nic.in/displaypdf.php?id=R mluYWxOYXRpb25hbFN1bW1hcnkucGRm

13. CPCB. Quality monitoring, emission inventory and source apportionment study for Indian cities. Central Pollution Control Board.https://cpcb.nic.in/source-apportionmen t-studies/

14. National Centre for Diseases Control. Impact of air pollution on health: A compendium of Indian studies. National program on climate change and human health. National Centre for Diseases Control. 2020. https://ncdc. mohfw.gov.in/wp-content/uploads/2024/04/580831283 21601281946.pdf

15. WHO EMRO. 9 out of 10 people worldwide breathe polluted air. Media centre. World Health Organization. https://www.emro.who.int/media/news/9-out-of-10-p eople-worldwide-breathe-polluted-air.html

16. WRI. Clean air for all: Financing clean air in India, report to the XVth Finance Commission, Government of India. WRI. _summary_report.pdf. https://fincomindia.nic.in/ archive/writereaddata/html_en_files/fincom15/StudyR eports/WRI_summary_report.pdf

17. Indian Council of Medical Research. First comprehensive estimates of the impact of air pollution on health loss and life expectancy reduction in each state of India. ICMR. 2018 Dec 06. 1702896411_press.pdf. https://www.icmr. gov.in/icmrobject/custom_data/1702896411_press.pdf

18. National Programme on Climate Change & Human Health. Health advisory on air pollution. Ministry of Health and Family Welfare. 2022 Sep. https://ncdc.mohfw.gov.in/w p-content/uploads/2024/04/3065716611669017053.pdf

19. Ministry of Health and Family Welfare. National action plan for climate change and human health. Ministry of Health and Family Welfare, Government of India. 2018 Oct 23. 27505481411548674558.pdf. https://ncdc.mohfw .gov.in/wp-content/uploads/2024/04/275054814115486 74558.pdf

20. Ministry of Health and Family Welfare. National action plan for climate change and human health. Ministry of Health and Family Welfare, Government of India. 2021 Feb. https://ncdc.mohfw.gov.in/wp-content/uploads/20 24/07/NATIONAL-ACTION-PLAN-FOR-CLIMATE-CHA NGE-HUMAN-HEALTH-2021.pdf

21. India State-Level Disease Burden Initiative Air Pollution Collaborators.   Health and economic impact of air pollution in the states of India: The Global Burden of Disease Study 2019. Lancet Planet Health. 2020 Dec 27; 5(1). - PMChttps://pmc.ncbi.nlm.nih.gov/articles/PMC7805008/

22. Basu J. National Clean Air Programme missed 2024 target to push back pollution, study shows. Down To Earth. 2024. https://www.downtoearth.org.in/pollution/national-clean-air-programme-missed-2024-target-to-push-back-pollution-study-shows-93848

23. Ganguly T, Selvaraj KL & Guttikunda SK. National Clean Air Programme (NCAP) for Indian cities: Review and outlook of clean air action plans. Atmospheric Environ X. 2020 Dec 1;8:100096. https://www.sciencedirect.com/science/article/pii/S2590162120300368

24. CREA. Tracing the hazy air 2024: Progress report on National Clean Air Programme (NCAP). Centre for Research on Energy and Clean Air. 2024 Jan 10. https://energyandcleanair.org/publication/tracing-the-hazy-air-2024-progress-report-on-national-clean-air-programme-ncap/

25. CREA. Tracing the hazy air 2024: Progress report on National Clean Air Programme (NCAP). Centre for Research on Energy and Clean Air. 2022 Jan. NCAP-Report_Jan22_.pdf. https://energyandcleanair.org/wp/wp-content/uploads/2022/01/NCAP-Report_Jan22_.pd

f

26. CREA. Tracing the hazy air 2024: Progress report on National Clean Air Programme (NCAP). Centre for Research on Energy and Clean Air. 2023. https://energya ndcleanair.org/wp/wp-content/uploads/2023/01/Tracin g-the-Hazy-Air-2023_-Progress-Report-on-National -Clean-Air-Programme-NCAP_10th-January-2023.pdf

27. CREA. Tracing the hazy air 2024: Progress report on National Clean Air Programme (NCAP). Centre for Research on Energy and Clean Air.

28. India State-Level Disease Burden Initiative Air Pollution Collaborators. The impact of air pollution on deaths, disease burden, and life expectancy across the states of India: The Global Burden of Disease Study 2017. The Lancet Planetary Health. 2019 Jan. https://www.thelanc et.com/journals/lanplh/article/PIIS2542-5196(18)30261 -4/fulltext

29. Centre for Science and Environment. Coping with climate change: An analysis of India's National Action Plan on Climate Change. 2018. https://cdn.cseindia.org/attachm ents/0.55359500_1519109483_coping-climate-change- NAPCC.pdf

30. Wang C & Cao J. Air pollution, health status and public awareness of environmental problems in China. Scientific Reports. 2024 Aug 27. https://www.nature.co m/articles/s41598-024-69992-2

31. National Centre for Disease Control. National Programme on Climate Change & Human Health. National Centre for Disease Control (NCDC). 2025 Jun 03.https://ncdc.mohfw.gov.in/national-programme-on-climate-change-human-health/

32. National Centre for Disease Control. Centre for Environmental & Occupational Health, Climate Change & Health. National Centre for Disease Control (NCDC). 2025 Jun 03. https://ncdc.mohfw.gov.in/centre-for-environmental-occupational-health-climate-change-health/

33. Crooks JL, Licker R, Hollis AL, & Ekwurzel B. The ozone climate penalty, NAAQS attainment, and health equity along the Colorado Front Range. J Expo Sci Environ Epidemiol. 2021 Sep 10. https://pmc.ncbi.nlm.nih.gov/articles/PMC9349035/

34. Wang C & Cao J. Air pollution, health status and public awareness of environmental problems in China. Scientific Reports. 2024 Aug 27. https://www.nature.com/articles/s41598-024-69992-2

35. Abhijith K, Kumar P, Omidvarborna H, Emygdio APM, McCallan B & Carpenter-Lomax D. Improving air pollution awareness of the general public through citizen science approach. Sustainable Horizons . 2024 Jun 1 ;10:100086. https://www.sciencedirect.com/science/article/pii/S2772737823000408

36. 36. National Programme on Climate Change and Human Health, National Centre for Disease Control, Directorate General of Health Services, Ministry of Health and Family Welfare, Government of India. Health adaptation plan for diseases due to air pollution. Ministry of Health and Family Welfare. 2021 Jul. https://ncdc.mohfw.gov.in/wp-content/uploads/2024/05/1.Health-Adaptation-Plan-for-Disease-Due-to-Air-Pollutions-2021.pdf

37. Global Climate & Health Alliance. Clean air, healthy lives: A policy roadmap for health systems to tackle air pollution. Global Climate & Health Alliance. 2025 Mar. https://climateandhealthalliance.org/wp-content/uploads/2025/03/Clean-Air-Healthy-Life-Report-1.pdf

38. National Programme on Climate Change and Human Health. Climate change and health: Driving local action. NCDC. 2023 Jul 12-14.https://ncdc.mohfw.gov.in/wp-content/uploads/2024/09/Report_National-Workshop_District-Capacity-Building-Case-Studies_2023.pdf

39. World Health Organization. Building health workforce capacity on air pollution and health: A train-the-tainer pilot workshop in Ghana. World Health Organization. https://iris.who.int/bitstream/handle/10665/373349/9789240077973-eng.pdf

7

Plastic Pollution Beyond Waste: A Systemic Approach to its Ecotoxicological and Health Impacts

Abstract

Plastic pollution has traditionally been perceived primarily as a waste management challenge. Until recently, the global discourse on plastic pollution focused on waste management overlooking its impact on public and marine health and ecosystem stability. This chapter delves into the far-reaching impacts of plastic pollution including the persistence of microplastics and the detrimental effects of their chemical leachates on human wellbeing and marine biodiversity. Microplastic chemical additives, plasticizers, and emulsifiers from plastic waste infiltrate food chains, water sources, and even the air we breathe instigating obvious and latent hazards such as endocrine disruption, neuropsychic and reproductive complications, genomic-based alterations, and diseases in humans. In marine environments, plastic debris constitutes a menace threatening aquatic species through ingestion and entanglement and alters water quality and oceanic biochemical processes ultimately impacting global climate regulation and upsetting marine biodiversity stability. The multidimensional effects of plastic pollution emphasize that there is a need for interdisciplinary solutions, policy interventions, and sustainable innovations to address the interconnectedness of ecological and health-related threats posed by plastic pollution.

Keywords: Plastic pollution, environmental toxicology, microplastics, biodiversity loss, public health, environmental policy, ocean plastic debris

Authors

Paul Ayomide Eweola, Founder and Executive Director, Aquaworld Community Development Initiative, Nigeria

Kehinde Adejugbagbe, Research Advisor, Aquaworld Community Development Initiative, Nigeria

Paul Olatunji, Foreign Hub Lead, Aquaworld Community Development Initiative, Nigeria

Mustapha Suebat, Member, Aquaworld Community Development Initiative, Nigeria

Simbiat Salaudeen, MSc Student, University of Basque Country, Spain

Adetola Adebowale, Assistant Project Manager, Aquaworld Community Development Initiative, Nigeria

Olagboye Olasunkanmi, Senior Programs Manager, Aquaworld Community Development Initiative, Nigeria

Ogunseitan Power, Research Associate, Aquaworld Community Development Initiative, Nigeria

Introduction

Overview of plastic pollution

Plastic pollution is one of the most pressing environmental challenges of the 21st century. It is due to the widespread use and persistence of plastic materials in the environment. Global plastic production exceeds 300 million metric tonnes annually with only a fraction being effectively recycled or incinerated with the majority ending up in landfills or in the natural environment and intensifies ecological degradation (1)(2). Plastics are composed of synthetic polymers designed for durability and are resistant to natural decay which results in their accumulation in terrestrial and aquatic ecosystems (3). Their persistence leads to serious environmental implications particularly through the formation of microplastics particles <5 mm that originate from the breakdown of larger plastics and other personal care products such as microbeads, dental floss, etc. Plastic particles and leachates or wastes have now permeated the earth's most remote corners from the polar ice caps to deep ocean trenches (4).

Plastics affect the global carbon cycle by altering plankton populations and primary productivity in aquatic and terrestrial systems. They directly threaten marine and terrestrial wildlife causing both lethal and sublethal harm to organisms ranging from whales and birds to corals and planktons (5). Microplastics serve as vectors for persistent organic pollutants (POPs) which accumulate in animal tissues and may pose unconfirmed but probable health risks to humans through ingestion via water and food (6). Soil ecosystems are not exempted. Plastic pollution disrupts soil health by altering pH levels, water retention, and microbial enzymatic activity, thereby hindering ecological processes and reducing soil fertility (7). Plastics release toxic pollutants into the environment, contributing to land, water, and air pollution and causing long-term harm due to their non-biodegradable nature (8).

Despite global efforts to curb plastic use, the scale of production and improper disposal practices continue to outpace regulatory interventions, suggesting that a fundamental transformation is expediently required in the 'plastic circular economy' model ranging from production methods to consumption behavior and waste management systems (9).

The objectives of this chapter are to: 1) examine the toxicological and biological impacts of plastic pollution; 2) highlight the applicable systemic and non-systemic nature of plastic pollution crisis; and 3) offer actionable and sustainable interventions and policies.

Shifting perspective

Global conversation around plastic pollution is undergoing a momentous transformation. Increasing awareness of its pervasive environmental impact has spurred a shift from passive observation to active intervention. Historically perceived as a symbol of modern convenience due to its durability and affordability, plastic is now widely recognized as a persistent environmental nuisance with far-reaching ecological and health consequences (3)(10). This evolving perspective is reflected in the mounting concerns over microplastics, which originate from the degradation of larger plastic items. These particles have infiltrated ecosystems worldwide and pose serious risks to marine life, soil quality, and human health (4)(7). The growing understanding that plastics alter global carbon cycles and interfere with the primary production in various ecosystems has increased the urgency of mitigating plastic-related threats (5).

The scale of global plastic production exceeding 300 million metric tonnes annually has intensified calls for an in-depth approach and efforts for holistic systemic changes (1). Efforts are increasingly aimed at reducing dependency on petrochemical-based plastics through innovation to biodegradable alternatives and policy reforms. However, the volume of plastic waste continues to rise (9)(2). Narratives surrounding plastic pollution are no longer confined to localized concerns, it has emerged as a transnational environmental crisis demanding collective reimagining of plastic chemicals and their variants approval, production, consumption, waste management practices, advocacy, and policy across nations.

Plastic and environmental interaction

Plastic pollution is now seen as a complex environmental problem that is not only prevalent as surface litter but is also an invisible menace that has far-reaching ecological consequences. One major issue related to plastic litter is that the larger items are broken down into microplastics and nanoplastics which interact with ecosystems in air, water, and soil. These interactions are mediated through a suite of physical, biological, and anthropogenic processes that control the transport and retention of plastics in the environment.

Microplastics in the environment

Microplastics (defined as plastic particles 5 mm or less in size) come from both primary sources (e.g., microbeads and industrial pellets) and secondary sources such as the degradation of larger plastics. Their small size enables them to be transported over long distances in the environment, and they can remain there for long periods, making them difficult to control in terms of ecosystem health and contaminant management. These varied and interlinked systems generate diverse pathways through which humans interact with the environment.

Pathways of dispersion: Air, water, soil

In the atmosphere, wind transports airborne particles across both terrestrial and aquatic regions (11). Within aquatic systems, ocean currents and thermohaline flows contribute to the movement of microplastics from coastal surfaces into deep-sea environments including submarines, canyons, and trenches (12). Terrestrial environments are affected by surface runoff which conveys plastic particles from inland soils to nearby freshwater and coastal systems.

In freshwater environments, microplastics often accumulate in lake and beach sediments influenced by sewage discharge and biological actions such as biofouling and settling (13). Human activities like textile shedding further contribute to the environmental microplastic load with marine shorelines showing over 85 percent of microplastic debris derived from synthetic microfibers (14).

Persistence and breakdown of plastics into microplastics and nanoplastics

Plastics exhibit remarkable environmental persistence resisting rapid degradation under natural conditions. Over time, physical forces e.g., ultraviolet radiation, wave action, chemical processes, and microbial activity gradually fragment larger plastics into microplastics and subsequently into nanoplastics—particles smaller than 1 μm (12). These smaller particles are more mobile, bioavailable, and chemically reactive, increasing their potential to infiltrate food chains and cause toxicological effects.

The persistence of microplastics is further enhanced by their resistance to biodegradation and their tendency to accumulate in environmental "sinks," such as deep-sea sediments and inland water bodies (13)(11).

Plastic chemical additives and toxic leachates

Endocrine-disrupting chemicals (EDCs) leaching from plastics pose significant health risks. Bisphenol A (BPA) and phthalates are associated with an increased risk of estrogen-dependent diseases in women including endometriosis and polycystic ovary syndrome (15). Persistent organic pollutants (POPs) and EDCs can impair neurodevelopment, immune function, and reproduction with particular concern of foetal, neonatal, and childhood exposure (16). Heavy metals like cadmium and lead are linked to endometriosis and endometrial cancer risk (15). Despite the benefits of plastics in modern society, their widespread use has raised concerns about their long-term health effects (17). These chemicals can leach from plastic containers into food and drinks, especially when heated. The increasing production and use of plastics and their potential to disrupt natural processes pose a significant challenge to human health and environmental sustainability (17).

Plastics and microplastics in trophic levels

The pervasive presence of plastics and microplastics (MPs) in aquatic ecosystems has triggered profound concern regarding their interactions across trophic levels. These synthetic pollutants, introduced primarily through anthropogenic activities, are dynamic vectors capable of transporting hazardous compounds including persistent organic pollutants (POPs), heavy metals, and pathogenic microorganisms. The ecological ramifications of these materials influence broader food web structures and processes.

Microplastic food chain interaction

Microplastics enter the marine food web mostly through lower trophic organisms such as phytoplankton and zooplankton which often mistake these particles for food due to their size and appearance. For instance, zooplankton, a fundamental dietary component for many secondary consumers, can ingest MPs and facilitate their upward movement through the food chain (18).

The ingestion of MPs by prey organisms often results in the formation of hetero-aggregates in the digestive tract, thereby delaying their clearance and increasing the likelihood of their transfer to predators (19). MPs also elicit physiological and biochemical responses including oxidative stress, impaired reproduction, and neurotoxicity (20). Their role as chemical vectors—adsorbing and transporting hazardous substances— amplifies their toxicological footprint within trophic networks (21).

Studies have confirmed trophic transmission from copepods to fish predators which further confirms the dual route through which microplastics permeate food webs— direct ingestion and indirect consumption via contaminated prey (22). The bioavailability and transferability of MPs is influenced by particle characteristics such as size, shape, and polymer composition (23). Fish inhabiting mid-water or surface regions, such as pelagic species, are disproportionately affected due to their ecological niches (24).

Accumulation in higher trophic levels (fish, marine mammals, and ultimately humans)

As MPs ascend the food chain, bioaccumulation and biomagnification become critical processes that intensify their ecological and health-related consequences. Evidence shows that fish at higher trophic levels such as *Esox lucius*, *Cyprinus carpio*, and *Luciobarbus caspius*- exhibit significant correlations between MP contamination and physiological traits like gill mass or age (25).

The contamination profiles of marine mammals also reflect cumulative exposure through complex trophic pathways which markedly differ from laboratory simulations due to metabolic variations and environmental diversity (26). The contamination of marine biota by MPs thus acts as a proxy for understanding the potential risks faced by humans.

Human exposure primarily occurs through the consumption of contaminated seafood. Recent studies suggest that MPs not only reach human diets but may also carry with them a suite of toxic substances that pose risks of endocrine disruption, carcinogenicity, and liver dysfunction (27). This is particularly concerning given the continuous release and breakdown of plastics in aquatic environments resulting in a persistent source of bioavailable contaminants (28)(29).

Interactions with environmental contaminants

Plastics, particularly microplastics (MPs), possess inherent physicochemical characteristics that enable them to interact extensively with environmental contaminants. Their high specific surface area, hydrophobicity, and polymeric structures make them ideal vectors for a wide range of hazardous pollutants including heavy metals, persistent organic pollutants (POPs), antibiotics, and pathogenic microorganisms. These interactions have amplified the ecological and toxicological implications of plastic pollution, particularly in aquatic and terrestrial ecosystems.

Plastics as carriers of pathogens and heavy metals

Microplastics in aquatic environments frequently act as substrates for microbial colonization forming dynamic biofilms known as the "*plastisphere.*" These biofilms facilitate the movement of pathogenic microorganisms including antibiotic-resistant bacteria (ARB) and genes (ARGs) across ecosystems, posing a critical threat to ecological and public health (30)(31). Plastics also serve as environmental reservoirs where horizontal gene transfer and co-selection mechanisms may enrich and disseminate resistance traits, aggravating the global health burden of antimicrobial resistance (32)(33).

In addition to pathogens, MPs also demonstrate a high propensity for adsorbing heavy metals such as lead (Pb), cadmium (Cd), chromium (Cr), and zinc (Zn). These interactions are governed by various factors including particle size, surface charge, zeta potential, and aging processes (34)(35). Environmental variables such as pH, salinity, and organic matter content further modulate the extent of metal sorption and mobility (36). Studies have shown that ageing increases the specific surface area and functional groups of MPs, enhancing their capacity to bind with heavy metals and facilitating their transport within soil and aquatic systems (36)(37).

These interactions increase the environmental persistence and bioavailability of toxic substances and raise the risk of biomagnification through food webs, ultimately threatening human and ecosystem health.

Role of microplastics in altering soil and water chemistry

Microplastics are increasingly recognized as agents that significantly modify soil and water chemistry. In terrestrial systems, especially agricultural soils, MPs interact with soil amendments, pollutants, and soil fauna, disrupting essential processes such as carbon and nitrogen cycling and altering microbial biodiversity and enzymatic functions (38)(39). These interactions often result in decreased soil fertility, modified nutrient dynamics, and impaired microbial community structure, compromising the multifunctionality of soil ecosystems (40).

Microplastic contamination influences key soil properties including pH, bulk density, electric conductivity, porosity, and water retention capacity. For instance, the presence of MPs can increase the mobility of certain heavy metals, rendering them more bioavailable and potentially more toxic to both plants and microorganisms (41). In aquatic environments, MPs can similarly alter the distribution and transport of chemical pollutants including pesticides, pharmaceuticals, and organic compounds by serving as sorptive carriers (42)(43).

MPs can modify aquatic chemical equilibria by affecting the pH, dissolved organic carbon content, and concentration of nutrients such as nitrogen and phosphorus (40). These changes influence primary productivity and biogeochemical cycles, thereby disturbing the ecological balance and functional integrity of freshwater and marine ecosystems.

Public health implications of plastic pollution

Plastics pollution poses a significant public health concern due to the intrinsic chemical properties of plastics and their ability to act as vectors of harmful chemicals (44)(45). Plastic polymers are composed of various persistent chemicals such as bisphenol A (BPA), phthalates, brominated flame retardants, and plasticizers which are added during the manufacturing process to enhance flexibility and durability (46). These substances are not chemically bound to plastics and are readily released in humans and within the environment (47). Many of these additives are considered priority toxic pollutants and deleterious even at low concentrations (48)(49). Microplastics also facilitate the transport of other harmful chemicals and microbes in humans and other biota further exacerbating their environmental and public health risks (50). Microplastics resistance to biodegradation, given their physicochemical descriptors of compositional chemical molecules such as high molecular weight, boiling and melting points, polarity, solubility (i.e. lipophilicity and hydrophobicity), etc. enhances the persistence of these toxic chemicals which leads to potential long-term negative impacts on biological activity and public health (51)(52). Unpredictably, humans are exposed to these plastic-based contaminants through various known and unknown pathways which make plastic pollution a significant maze for public health.

Human exposure pathways

Humans are exposed to microplastics primarily through ingestion, inhalation, and, less often, via dermal contact with ingestion of contaminated food items being the predominant route (23)(53). Common dietary items such as fruits and vegetables, milk and honey, table salt, and food packaging materials have been reported as sources of microplastic exposure to humans (Figure 1) (54)(55)(56)(57)(58). Even the act of repeatedly opening and closing plastic bottles can contribute to microplastic contamination (59). Applying sewage sludge as fertilizers in agricultural fields may increase microplastics uptake by crops (60).

An average person is estimated to consume about 46,000 microplastic particles annually (61). Seafood consumption presents a unique pathway of microplastic ingestion in humans (62)(63). Aquatic organisms, particularly fish, which are known to be hyper converters of food or feed whether in the wild or under culture conditions, would possibly bioaccumulate and biomagnify microplastics in the aquatic food web. These are not readily excreted due to slow or incomplete elimination or high gut retention within the gastro-intestinal tract and other tissues of fish (64). Shellfish such as mussels are especially vulnerable due to their filter-feeding mechanisms, which allow them to absorb microplastic particles in their gills (65). This slow or incomplete elimination of plastics in fish predisposes humans and other dependents on fish for nutrition and other industrial applications to plastic pollution. Currently, there are no regulations for microplastics in seafood (63). Inhalation of airborne microplastics, particularly in urban and industrialized cities, is another microplastic exposure pathway in humans. Air studies have found high microplastic levels in humans due to industrial and vehicular emissions (66). A study in Paris measured up to 200 microparticles/m^3/day in air (67). Inhalation of microplastics and their additives can also occur in indoor environments via household dust and synthetic fabrics such as nylon or polyester which are used in clothing, carpets, and electronics (68).

Figure 1

Potential pathways of microplastics and their maximum abundance observed for ingestion or inhalation to human beings

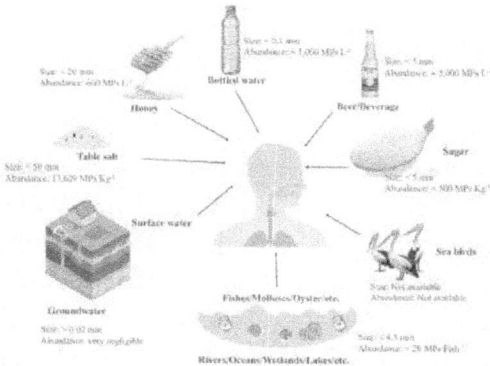

Source: The potential of microplastics as carriers of metals. Environmental Pollution, 255, 113363 (42).

Plastics as endocrine and reproduction disruptors

Plastic chemicals and additives negatively impact fertility, hormonal balance, and fetal development (69). For example, endocrine disruptors such as BPA and phthalates mimic natural hormones and contribute to health conditions such as diabetes, hormonal imbalance, and obesity (70). These compounds can infiltrate the blood-placental barrier and induce neurodevelopmental abnormalities in fetus (71). Pregnant women with high levels of microplastic pollutants could experience maternal-fetal membrane disruption, where the protective barrier between the mother and in the fetus is compromised paving the way for incursion of harmful substances alongside beneficial nutrients and gases. Increasing evidence suggests that in-utero microplastic exposure may adversely affect fetal and neonatal body weight and organ development, while also disrupting the normal release of gonadotropins and reproductive hormones, thereby impairing human reproductive development (72)(73).

Carcinogenicity and chronic diseases

Most plastic additives are potentially carcinogens (74). Long-term exposure to these chemicals has been linked to mitochondrial dysfunction and oxidative stress which led to metabolic disorders, cardiovascular diseases, and impaired neuronal development (75)(76). Human and other model animal studies have linked long-term microplastic exposure to microplastic chemicals with an increased incidence of tumors including breast and liver cancers, neurodegenerative diseases, chronic inflammation, and compromised intestinal barrier function especially in rats (77)(78)(79)(80)(81).

Respiratory and immune system disorders

Inhaling airborne microplastic fibers can severely impact respiratory and immune health (82). Once inhaled, these fibers are sequestered within the lung lining where they trigger inflammatory and oxidative processes that can cause lung damage (83). They are not easily eliminated by the body's defense mechanism (84). High microplastic exposure levels have been linked to chronic respiratory diseases such as chronic obstructive pulmonary disease and pulmonary fibrosis while also compromising immune cells and disrupting immune-related signaling pathways (85)(63)(86)(87). Residents and workers in industrial areas may be more prone to the respiratory effects of microplastics (88). For instance, a high amount of synthetic microplastic fibers were found in exposed occupational workers which could increase respiratory irritation and lesion development in the lungs (89). Furthermore, high phthalate levels are correlated with the occurrence of asthma in both children and adults (90)(91). These findings highlight the potential that microplastics have to contribute to lung inflammation, asthma, and weakened immune responses.

Marine ecosystem disruptions due to plastic pollution

The consequences of plastic pollution go beyond waste management concerns as it impacts various ecosystems. In marine ecosystems, plastic pollution causes physical harm and complex biochemical disruptions, making it a top environmental threat.

Plastic ingestion and its impact on marine life

Marine organisms are increasingly ingesting plastics in their environment, posing threats to their health and survival. Studies have shown an increase in the quantity of plastic debris in the diets of marine species. Species such as sea turtles have been found with stomachs full of plastics (92). For example, loggerhead sea turtles have been observed to ingest plastics while mistaking them for jellyfish leading to malnutrition, intestinal blockages, and death in severe cases (93).

The bioaccumulation of toxic microplastics is another concerning issue affecting both marine biodiversity and human health. Toxic chemicals such as heavy metals and persistent organic pollutants (POPs) can be embedded in microplastics. When ingested by marine organisms, these microplastics can enter the food chain and accumulate in large predators. Also, chemicals embedded in microplastics can leach into marine organisms, causing long-term ecotoxicological damage (94).

Studies reveal that plastic pollution is directly linked to population declines in flesh-footed shearwater. It is estimated that 99 percent of seabird species may be impacted by 2050; 69 percent and 74 percent Hawaii and Northeast Atlantic seabirds had ingested plastics, respectively (95)(96)(97). Uncontrovertibly, chemical leachates of plastic origin have been reported in seabird tissues (98)(99).

Microplastics were found in all 102 sea turtles comprising

seven species worldwide indicating exposure due to contamination of saltwater, sediments, and diets (100). Twenty two of thirty sperm whales stranded along the coast of the North Sea had consumed plastic materials such as nets, bags, and packaging material from industrial products (101). Also, microplastic poisoning has been linked to fish mortality along with other processes like physiological perturbations and organ infections. The European Sea bass larvae had growth impairment due to exposure to higher concentrations of polyethylene microbeads (102).

These chemicals can also affect human health when contaminated seafood is consumed potentially leading to endocrine disruption, neurotic complications, and reproductive disorders (103).

Figure 2

The effects of microplastic and pollutants in marine organisms

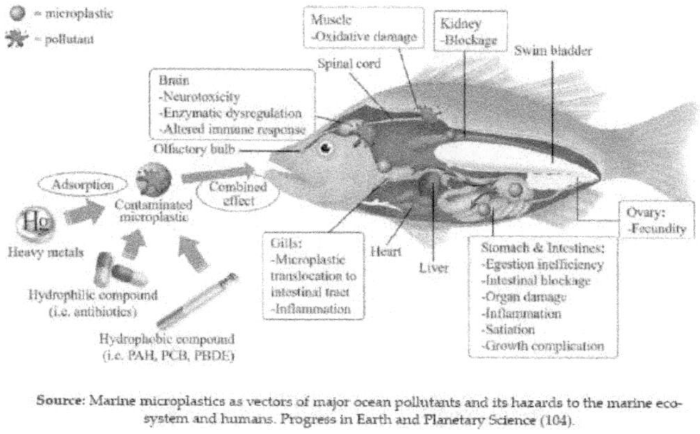

Source: Marine microplastics as vectors of major ocean pollutants and its hazards to the marine ecosystem and humans. Progress in Earth and Planetary Science (104).

265

Physical hazards of plastic waste

Aside from ingestion, plastic waste also poses physical hazards to marine organisms particularly through entanglement and suffocation. A study estimated that more than 700 marine species are impacted by plastic waste and that many of these species are critically endangered (105). Entanglement with plastic waste like fishing nets and plastic bags can cause drowning or starvation as entangled animals are not able to hunt for food, and escape predators due to the constraints of plastic waste. For example, 81 percent of kelp gulls in Patagonia died after being trapped by a fishing line (106). Plastic ingestion impacts on marine organisms such as whales and even planktons have been documented. Plastics cause physiological damage, oxidative stress, and neurotoxicity (107)(108). These hazards compound the biological challenges marine ecosystems face such as overfishing, habitat destruction, and climate change.

Figure 3
Fate of plastic waste showing entanglement

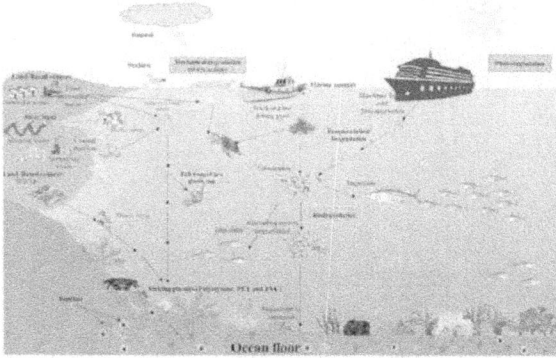

Source: Insightful analytical review of potential impacts of microplastic pollution on coastal and marine ecosystem services (109).

Alteration of ocean biogeochemistry

Plastics that settle on surface waters interact with microorganisms. In this process, they alter the microbial communities involved in nutrient cycling and carbon fixation and displace the equilibrium of oxygen-producing phytoplankton and CO_2-releasing bacteria inhibiting photosynthesis and influencing oceanic carbon balance (110)(111).

Also, plastics can serve as a substrate for the formation of biofilms which enhance particle density causing microplastics to sink and disrupt the balance of marine ecosystems. These alterations can affect the ocean's regulation of climate by reducing carbon sequestration (112). Plastics provide a haven for harmful bacteria to negatively impact the health of marine lives.

Plastics ingested by phytoplankton decrease carbon fixation, promote ocean acidification which disrupts carbon sinks, elevating CO_2 levels. This results in plastic pollution, climate change, and ecosystem collapse (113). In seabeds, microplastic deposits limit oxygen diffusion enabling anaerobic metabolism and suppressing carbon sequestration (114).

Figure 4
Impacts of plastic pollution on marine carbon biogeochemistry

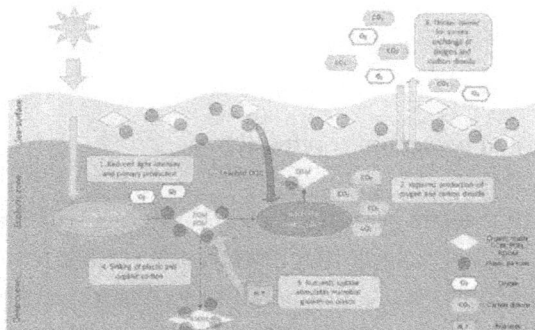

Source: Plastic pollution impacts marine carbon biogeochemistry (115).

Plastics have made a significant contribution to greenhouse gas production and breakdown with the potential of contributing 13 percent of the earth's carbon budget in 2050 (116). Long-lasting plastic marine debris emits greenhouse gases enhancing climate feedback loops. Ocean acidification due to rising CO_2 intensifies the toxicity of microplastics by influencing chemical adsorption and disturbing the marine food web (114).

268

As plastics degrade on shellfish and interact with seawater, they release chemicals, such as bisphenol A (BPA) and phthalates, which can alter the buffering capacity of marine ecosystems, contributing to ocean acidification which in turn, disrupts marine life, especially shell-forming organisms such as mollusks and corals (94).

Loss of biodiversity and ecosystem services

The economic impacts of plastic pollution are far-reaching. Reduction in fisheries' catch is affecting the market supply of the affected species. The unpleasant sight of polluted beaches and damaged coral reefs has caused a reduction in tourism in some coastal regions leading to revenue loss for local economies (117). Since the main source of protein for over 1.4 billion people depends on seafood, fish populations face a humongous decline due to environmental pollution chiefly attributed by plastics pollution via ingestion, entanglement, and chemical interactions of plastic chemical leachates. Plastics in the ocean contributed $7 billion of GDP losses in 2018 through reduced fishery revenues linked to equipment damage, reduced catches, and tourism losses due to polluted beaches (118).

Cleanup and repairs of equipment are direct costs and biodiversity loss, and degraded habitats are indirect costs of plastic pollution. If not curbed, plastic pollution will continue to double annually and is projected to reach 44 million tonnes by 2060 with economic and food security at high risk (119).

Critical habitats, such as coral reefs and mangroves, have been disrupted by plastic pollution resulting in a loss of biodiversity and ecosystem services. Coral reefs are particularly vulnerable to the physical damage caused by plastic debris affecting their ability to provide food, shelter, and breeding grounds for marine species. Also, coral polyps can be suffocated by plastics reducing their ability to feed and reproduce. This can eventually lead to coral bleaching (120).

Similarly, mangroves and seagrass beds are increasingly threatened by plastic waste through suffocation and degradation, reducing the efficiency of their ecological functions.

Policy interventions and systemic solutions

Regulatory frameworks and international efforts

The problem of plastic pollution is transboundary, crossing beyond the national jurisdiction of each country (121)(122). Hence, there is a need for international policies and other interventions to curtail the impact of this global menace (123)(124).

Several global treaties and agreements have been formulated to combat plastic pollution. Efforts made by governments and international organizations, particularly through national, regional, and international action plans, initiatives and instruments including relevant multilateral agreements are welcome. Various international regulatory frameworks have been in place to address plastic pollution. The United Nations Convention on the Law of Sea (UNCLOS), established in 1982 with a marine pollution legal framework is a mandate for member states (121)(123)(124). However, UNCLOS, only compels member states to have domestic regulations and laws to mitigate and monitor marine pollution within their own jurisdiction. Enforcing prevention, reduction, and control of plastic waste from reaching the oceans is left to member states (124).The Basel Convention alongside the Declaration of Alliance of Small Island States were established to address transboundary chemical and hazardous waste disposal to ensure responsible plastic waste recycling and to reduce pollution in developing ocean, coastal island states, and countries (125)(124). However, it was not until the convention's fourteenth meeting in May 2019 that the conference parties included efforts to prevent plastic waste from leaking into the ocean, especially from the plastic waste exported to developing countries (126).

The United Nations (UN) Plastics Treaty aims to regulate plastic production and to mitigate waste management to forestall environmental and health hazards (127). And the Stockholm Convention regulates persistent organic pollutants (POPs) in plastics that lead to toxic contamination in ecosystems.

Another international effort aimed at reducing plastic pollution is the International Convention for the Prevention of Pollution from Ships (MARPOL) established by the International Maritime Organization (IMO) in 1978 to prevent inappropriate ship waste disposal (128). Most importantly, the MARPOL Annex V, amended in January 2013, prohibits solid waste disposal including plastics into the sea (129)(130)(131). Current international frameworks are not effective in combating aquatic plastic pollution. But the newly adopted resolution of the Fifth

Session of the United Nations Environment Assembly (UNEA-5.2) to initiate an international law-making process to combat global plastic pollution initially by 2024 (UNEP Resolution 5/14) looks promising (132). However, this conversation is still ongoing.

Regional and national policies on plastic ban and extended producer responsibility (EPR)

Different countries and regions have various regulations regarding plastic bans. These include the Osaka Blue Ocean Vision, the Ocean Plastics Charter; the Association of Southeast Asian Nations (ASEAN) Framework of Action which focuses on marine debris, the Bangkok Declaration on Combating Marine Debris in the ASEAN region, and the Asia-Pacific Economic Cooperation Roadmap on Marine Debris. The European Union implemented Directive (EU) 2019/904, referred to as the Single-Use Plastics Directive (SUPD) bans the use of 10 most used single-use plastic products. It also establishes further regulatory measures including extended producer responsibility (EPR) and some product design modifications (133)(134)(135). However, the EU leaves the member states with the prerogative to implement this directive to achieve a reasonable and measurable reduction of plastic pollution (135). While several countries including Luxembourg, Greece, France, Sweden, Slovenia, Cyprus, Latvia, and Denmark, have effectively implemented Directive (EU) 2019/904, others, such as the Czech Republic, Croatia, and Romania, have demonstrated limited commitment. Additionally, nations like Germany and the Netherlands have overlooked the key provisions of the directive highlighting significant disparities in its implementation across EU member states (136)(137). The directive also incorporates the provision of the Extended Producer Responsibility (EPR) frameworks which shifts attention and responsibility for plastic waste management from the consumers to the producers encouraging them to design products that are more eco-friendly and recyclable, leading to less plastic pollution (138)(139). Several states in the United States of America have also adopted EPR policies alongside additional measures like restrictions and bans on single-use plastics since the USA does not have a comprehensive federal law addressing this issue (140)(141).

Several other countries like Rwanda, Kenya, Bangladesh, Paraguay, Algeria, Cambodia, China, Botswana, Sri Lanka, and Burkina Faso, have formulated a range of policies to address plastic pollution including outright ban imposition of levies on plastic distribution at the retail level and EPR frameworks (142).

Circular economy and sustainable alternatives

Transitioning to a circular economy is one of the sustainable ways proposed to reduce plastic pollution in the environment (133)(143). This involves implementing the 7Rs model for plastic waste management (Figure 5). However, attention has been on the recycling part of the 7Rs whereby plastic waste is being converted into raw materials for other plastic products (144). Plastics can undergo mechanical recycling, dissolution-based approaches, pyrolysis/liquefaction, and hydrogenolysis depending on the polymer types (145).

Furthermore, bioplastics have been lauded as an eco-friendlier and more sustainable substitute for petrochemical plastics (146)(147). Bioplastics made from agricultural or bio-based products like corn, sugarcane, and vegetable oil became an alternative because they are more readily biodegradable compared to conventional plastics. However, research has shown that bioplastics production often involves repurposing productive land intended for food cultivation for the cultivation of raw materials for bioplastics and that bioplastics release methane in the atmosphere causing more ozone depletion than conventional petrochemical plastics (148)(149)(150).

Corporate responsibility and consumer behavior

Research on plastic pollution and its impacts on aquatic environments continues to expand together with the development of national, regional, and international policies to address this global challenge. However, effective and sustainable solutions require integrating scientific research into policy-making processes for improved plastic waste management and pollution reduction (143). There is a critical need to standardize methodologies used in plastic pollution studies to ensure that research outputs are comparable and capable of informing evidence-based policy (151). There should also be more investment to study the ecotoxicological impacts of plastics. Enhancing public education and awareness is essential for promoting responsible disposal practices and encouraging the adoption of more sustainable plastic alternatives (128)(152).

Figure 5
Circular economy principles for plastic waste
management

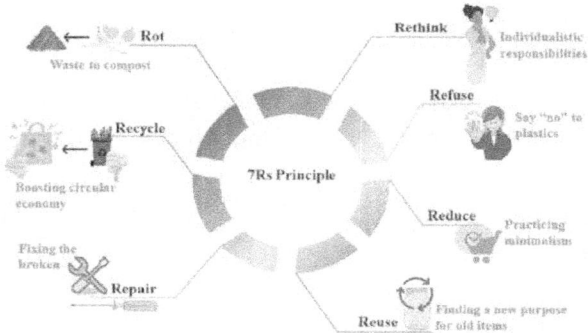

Source: Borah and Kumar 2024 (158).

Advancing research and scientific monitoring

Need for standardized microplastic detection methods

Microplastic pollution poses significant ecological and health risks necessitating standardized detection methods for accurate assessment. Techniques such as Fourier Transform Infrared Spectroscopy (FTIR) and Raman Spectroscopy that facilitates microplastic identification in environmental samples (153) are currently deployed alongside existing methodologies in microplastic detection.

The representativeness and reproducibility of most results are uncertain as they have poor methodology descriptions lacking important details such as the volume of the bulk sample. Thus, there is an urgent need for a standardized, quick and simple methodology. This new sampling protocol should include: (a) measures to reduce cross-contamination; (b) how and where to collect bulk samples; (c) how to separate microplastics from bulk samples, possibly through direct filtration, setting a filter's pore size, or with a previous exposure to a salt saturated solution such as NaCl; (d) a digestion protocol that is quick and has little effect on polymer integrity possibly H_2O_2 or enzymes; and (e) criteria for visual identification with the aid of staining dyes and recommended methods of chemical characterization (154)

Investing in environmental toxicology research

Research on environmental toxicology is essential for understanding the long-term health effects of plastic-related pollutants. Studies on endocrine-disrupting chemicals (EDCs) and microplastic bioaccumulation contribute to assessing potential health risks and formulating mitigation strategies (155). In 2024, 6.92 million plastic wastes were released into rivers, lakes, and seas (153) leading to an estimated monetary loss of up to USD 2.5 trillion in marine natural capital (156).

The issue of plastic waste in research laboratories is not just an environmental concern, it is a call for a systemic shift in how scientific research is conducted. Transition to sustainable laboratory practice requires dedication, funding, and policy changes. But the benefits far outweigh the challenges of driving a systemic shift toward sustainable research practices and a greener future. Scientists must lead by example ensuring that research progress does not come at the cost of environmental pollution (157).

Figure 6

Schematic illustration of the various kinds of plastics used in laboratories along with recycling strategies

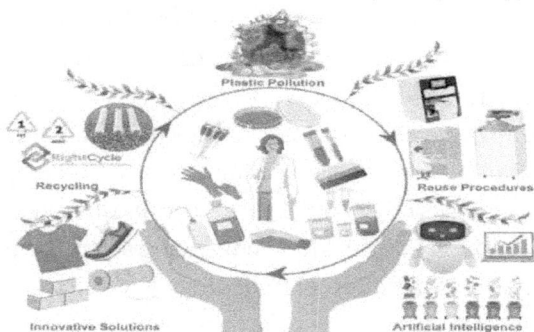

Source: Anil and Veda 2024

Reduce, reuse, and recycle is the transition needed: to reduce waste, maintain a proper inventory, utilize AI tools for sorting, and prevent bulk procurement; to reuse sterilize plastics using UV, autoclaving or chemical treatment; and recycle by transforming them into valuable products such as pellets, bricks, and clothes.

Concluding comments

Plastic pollution has evolved from a simplistic waste manage-ment issue into a complex, multidimensional crisis threat-ening ecological integrity, human health, and global sustain-ability. In this chapter, the authors have demonstrated how plastics, particularly microplastics and their toxic additives, permeate environmental systems, alter biogeochemical processes, and pose severe risks to both marine biodiversity and public health. The pervasive nature of plastic pollutants, their persistence in ecosystems, and their capacity to bioaccumulate across trophic levels demand urgent systemic interventions.

Addressing this crisis requires more than piecemeal solutions. It calls for a paradigm shift towards an interdisciplinary integrated framework involving science, policy, innovation, and societal behavior. Circular economy models, international policy frameworks, corporate accountability, and empowered consumer choices must converge to disrupt the current linear plastic economy. Furthermore, advancing scientific research, especially in environmental toxicology and microplastic detection, remains crucial for informing evidence-based policy and sustainable innovation.

As the global community moves forward, its priority must shift towards reducing plastic production at its source, investing in safe alternatives, enforcing producer responsibility, and fostering a culture of environmental stewardship. Only through collective systemic efforts can we mitigate the cascading impacts of plastic pollution and safeguard planetary and human health for future generations.

References

1. Nizzetto, L., & Sinha, S. Top priority to curb plastic pollution: Empowering those at the bottom. One Earth, 2(1), 11–15. 2020. https://doi.org/10.1016/j.oneear.2020. 01.005

2. Adeniran, A. O., Funmilayo Mary Ilugbami, & Gbemileke Tobi Oyeniran. A literature review on the effect of plastic waste deposits on soil ecosystem. Annals of Ecology and Environmental Science, 6(1), 23–31.2024. https://doi.org/10.22259/2637-5338.0601003

3. Chauhan, G. S., & Wani, S. Plastic pollution: a major environmental threat. International Journal of Innovative Research in Technology, 6(6), 43-46. 2019. https://www.researchgate.net/publication/341459854_Plastic_Pollution_A_Major_Environmental_Threat#references

4. Parmar, R. B., Agrawal, P. H., Rebuma, T., & Pal, M. Current Scenario on the Impact of Microplastics on the Environment, Marine, and Humans. OAJRC Environmental Science, 5(2), 34–38. 2024. https://doi.org/10.26855/oajrces.2024.06.001

5. Baranova, A., Novozhilova, T., Litovka, A., & Bilousov, M. The problem of plastic waste pollution in the world ocean. Bulletin of the National Technical University "KhPI" Series: New Solutions in Modern Technologies, 2(12), 69–73. 2022. https://doi.org/10.20998/2413-4295.2022.02.10

6. Rhodes, C. J. Plastic pollution and potential solutions. Science Progress, 101(3), 207–260. 2018. https://doi.org/10.3184/003685018x15294876706211

7. Rai, M., Pant, G., Pant, K., Becky Nancy Aloo, Kumar,

G., Singh, H. B., & Tripathi, V. Microplastic pollution in terrestrial ecosystems and its interaction with other soil pollutants: a potential threat to soil ecosystem sustainability. Resources, 12(6), 67–67. 2023. https://d oi.org/10.3390/resources12060067

8. Kumar, P. Impact of plastic on the environment. International Journal of Trend in Scientific Research and Development, 2(2), 471–474. 2018. https://doi.org/ 10.31142/ijtsrd9421

9. Kwon, G., Cho, D.-W., Park, J., Bhatnagar, A., & Song, H. A review of plastic pollution and their treatment technology: A circular economy platform by thermochemical pathway. Chemical Engineering Journal, 464, 142771–142771. 2023. https://doi.org/ 10.1016/j.cej.2023.142771

10. Selvamurugan Muthusamy, M., & Pramasivam, S. Bio-plastics – an eco-friendly alternative to petrochemical plastics. Current World Environment, 14(1), 49–59. 2019 https://doi.org/10.12944/cwe.14.1.07

11. Koutnik, V. S., Leonard, J., Alkidim, S., DePrima, F. J., Ravi, S., Hoek, E. M. V., & Mohanty, S. K. Distribution of microplastics in soil and freshwater environments: Global analysis and framework for transport modeling. Environmental Pollution, 274, 116552. 2021. https://doi. org/10.1016/j.envpol.2021.116552

12. Kane, I. A., & Clare, M. A. Dispersion, accumulation,

and the ultimate fate of microplastics in deep-marine environments: A review and future directions. Frontiers in Earth Science, 7. 2019. https://doi.org/10.3389/feart.2019.00080

13. Imhof, H. K., Ivleva, N. P., Schmid, J., Niessner, R., & Laforsch, C. Contamination of beach sediments of a subalpine lake with microplastic particles. Current Biology, 23(19), R867–R868. 2013. https://doi.org/10.1016/j.cub.2013.09.001

14. Carr, S. A. Sources and dispersive modes of micro-fibers in the environment. Integrated Environmental Assessment and Management, 13(3), 466–469. 2017. https://doi.org/10.1002/ieam.1916

15. Chitakwa, N., Alqudaimi, M., Sultan, M., & Wu, D. Plastic-related endocrine disrupting chemicals significantly related to the increased risk of estrogen-dependent diseases in women. Environmental Research, 252, 118966. 2024. https://doi.org/10.1016/j.envres.2024.118966

16. Damstra, T. Potential effects of certain persistent organic pollutants and endocrine disrupting chemicals on the health of children. Journal of Toxicology: Clinical Toxicology, 40(4), 457–465. 2002. https://doi.org/10.1081/clt-120006748

17. Halden, R.U. Plastics and health risks. Annual Review of Public Health. 31(1): 179–194. 2010. https://doi.org/

10.1146/annurev.publhealth.012809.103714

18. Botterell, Z. L. R., Beaumont, N., Dorrington, T., Steinke, M., Thompson, R. C., & Lindeque, P. K. Bioavailability and effects of microplastics on marine zooplankton: A review. Environmental Pollution. 245(245), 98–110. 2019. https://doi.org/10.1016/j.envpol.2018.10.065

19. Egbeocha, C., Malek, S., Emenike, C., & Milow, P. Feasting on microplastics: Ingestion by and effects on marine organisms. Aquatic Biology. 27, 93–106. 2018. https://doi.org/10.3354/ab00701

20. Iheanacho, S., Ogbu, M., Bhuyan, M. S., & Ogunji, J. Microplastic pollution: An emerging contaminant in aquaculture. Aquaculture and Fisheries.2023. https://doi.org/10.1016/j.aaf.2023.01.007

21. Seltenrich, N. New link in the food chain? Marine plastic pollution and seafood safety. Environmental Health Perspectives, 123(2). 2015. https://doi.org/10.1289/ehp.123-a34

22. McHale, M. E., & Sheehan, K. L. Bioaccumulation, transfer, and impacts of microplastics in aquatic food chains. Journal of Environmental Exposure Assessment, 3(3). 2024. https://doi.org/10.20517/jeea.2023.49

23. Nawab A, Ahmad M, Khan MT, Nafees M, Khan I & Ihsanullah I. Human exposure to microplastics: A review on exposure routes and public health impacts. Journal

of Hazardous Materials Advances. 16:100487 -10. 2024.
https://doi.org/10.1016/j.hazadv.2024.100487

24. Cruz, A. H. Impact of plastic waste ingestion by fish.
Circular Economy and Sustainability. Springer Nature.
2022. https://doi.org/10.1007/s43615-022-00242-1

25. Saemi-Komsari, M., Esmaeili, H. R., Behnam Keshavarzi,
Busquets, R., Abbasi, K., Farideh Amini Birami, &
AmirHassan Masoumi. Trophic transfer, bioaccumulation and translocation of microplastics in an
international listed wetland on the montreux record.
Environmental Research. 257; 119172–119172. 2024.
https://doi.org/10.1016/j.envres.2024.119172

26. Ross, P. S. Marine mammals as sentinels in ecological
risk assessment. Human and Ecological Risk Assessment:
An International Journal, 6(1), 29–46. 2000. https://doi.
org/10.1080/10807030091124437

27. Lehel, J., & Murphy, S. Microplastics in the food chain:
Food safety and environmental aspects. Reviews of
Environmental Contamination and Toxicology. 259,
1–49. 2021. https://doi.org/10.1007/398_2021_77

28. Batra, T. Endocrine - disrupting carcinogenic plastic
contamination in the food chain: A review. Biosciences
Biotechnology Research Asia, 8(2), 597–601. 2011.
https://doi.org/10.13005/bbra/905

29. Bisht, V. S., & Negi, D. Microplastics in aquatic ecosystem:

Sources, trophic transfer and implications. Int. Journal of Fisheries and Aquatic Studies. 8, 227-234. 2020. https://www.fisheriesjournal.com/archives/2020/vol8issue3/PartC/8-3-8-773.pdf

30. Sooriyakumar, P., Bolan, N., Kumar, M., Singh, L., Yu, Y., Li, Y., Weralupitiya, C., Vithanage, M., Ramanayaka, S., Sarkar, B., Wang, F., Gleeson, D. B., Zhang, D., Kirkham, M. B., Rinklebe, J., & M Siddique, K. H. Biofilm formation and its implications on the properties and fate of microplastics in aquatic environments: A review. Journal of Hazardous Materials Advances, 6, 100077. 2022. https://doi.org/10.1016/j.hazadv.2022.100077

31. Junaid, M., Liu, X., Wu, Y., & Wang, J. Selective enrichment of antibiotic resistome and bacterial pathogens by aquatic microplastics. Journal of Hazardous Materials Advances, 7, 100106. 2022. https://doi.org/10.1016/j.hazadv.2022.100106

32. Zhao, H., Hong, X., Chai, J., Wan, B., Zhao, K., Han, C., Zhang, W., & Huan, H. Interaction between microplastics and pathogens in subsurface system: What we know so far. Water, 16(3), 499–499. 2024. https://doi.org/10.3390/w16030499

33. Imran, Md., Das, K. R., & Naik, M. M. Co-selection of multi-antibiotic resistance in bacterial pathogens in metal and microplastic contaminated environments: An emerging health threat. Chemosphere, 215, 846–857. 2019. https://doi.org/10.1016/j.chemosphere.2018.10.11

4

34. Ding, T., Wei, L., Hou, Z., Li, J., Zhang, C., & Lin, D. Microplastics altered contaminant behavior and toxicity in natural waters. Journal of Hazardous Materials, 425, 127908–127908. 2022. https://doi.org/10.1016/j.jhazmat.2021.127908

35. Özgenç, E. Evaluation of the spreading dynamics and interactions of lead-carrier microplastics affected by biofilm: A mini-review. Water, Air, & Soil Pollution, 235(5). 2024. https://doi.org/10.1007/s11270-024-07090-9

36. Luo, H., Tu, C., He, D., Zhang, A., Sun, J., Li, J., Xu, J., & Pan, X. Interactions between microplastics and contaminants: A review focusing on the effect of aging process. Science of the Total Environment, 899, 165615–165615. 2023. https://doi.org/10.1016/j.scitotenv.2023.165615

37. Medyńska-Juraszek, A., & Jadhav, B. Influence of different microplastic forms on pH and mobility of cu2+ and pb2+ in soil. Molecules (Basel, Switzerland), 27(5), 1744. 2022. https://doi.org/10.3390/molecules27051744

38. Fan, C., Li, Y., Tian, C., & Li, Z. Effects of microplastics on soil C and N cycling with or without interactions with soil amendments or soil fauna. European Journal of Soil Science, 75(1). 2023. https://doi.org/10.1111/ejss.13446

39. Bouaicha, O., Mimmo, T., Tiziani, R., Praeg, N., Polidori, C., Lucini, L., Vigani, G., Terzano, R., Sanchez-Hernandez, J. C., Illmer, P., Cesco, S., & Borruso, L. Microplastics make their way into the soil and rhizosphere: A review of the ecological consequences. Rhizosphere, 22, 100542. 2022. https://doi.org/10.1016/j.rhisph.2022.100542

40. Feng, X., Wang, Q., Sun, Y., Zhang, S., & Wang, F. Microplastics change soil properties, heavy metal availability and bacterial community in a pb-zn-contaminated soil. Journal of Hazardous Materials, 424, 127364. 2022. https://doi.org/10.1016/j.jhazmat.2021.127364

41. Chen, L., Han, L., Feng, Y., He, J., & Xing, B. Soil structures and immobilization of typical contaminants in soils in response to diverse microplastics. Journal of Hazardous Materials, 438, 129555. 2022. https://doi.org/10.1016/j.jhazmat.2022.129555

42. Godoy, V., Blázquez, G., Calero, M., Quesada, L., & Martín-Lara, M. A. The potential of microplastics as carriers of metals. Environmental Pollution, 255, 113363. 2019 https://doi.org/10.1016/j.envpol.2019.113363

43. Kinigopoulou, V., Pashalidis, I., Kalderis, D., & Anastopoulos, I. Microplastics as carriers of inorganic and organic contaminants in the environment: A review of recent progress. Journal of Molecular Liquids, 350,

118580. 2022. https://doi.org/10.1016/j.molliq.2022.118
580

44. Ebrahimi P, Abbasi S, Pashaei R, Bogusz A & Oleszczuk
P Investigating impact of physicochemical properties
of microplastics on human health: A short bibliometric
analysis and review. Chemosphere. 289:133146. 2022.
https://doi.org/10.1016/j.chemosphere.2021.133146

45. Stapleton MJ & Hai FI. Microplastics as an emerging
contaminant of concern to our environment: A brief
overview of the sources and implications. Bioengineered.
2023 Aug 8;14(1). https://doi.org/10.1080/21655979.202
3.2244754

46. Hahladakis JN, Velis CA, Weber R, Iacovidou E & Purnell
P. An overview of chemical additives present in plastics:
Migration, release, fate and environmental impact
during their use, disposal and recycling. Journal of
Hazardous Materials. 344(344):179–99. 2018. https://d
oi.org/10.1016/j.jhazmat.2017.10.014

47. Blackburn K & Green D. The potential effects of
microplastics on human health: What is known and
what is unknown. Ambio. 51(3):518–30. 2021. https://do
i.org/10.1007/s13280-021-01589-9

48. Bucci K, Tulio M & Rochman. What is known and
unknown about the effects of plastic pollution: A
meta-analysis and systematic review. Ecological
Applications. 30(2). 2020. https://doi.org/10.1002/e

ap.2044

49. Azoulay D, Villa P, Arellano Y, Gordon M, Moon D, Miller K & Thompson K. Plastics and health the hidden cost of a plastic planet. Center For Environmental Law. 2019. https://www.ciel.org/wp-content/uploads/2019/02/Plastic-and-Health-The-Hidden-Costs-of-a-Plastic-Planet-February-2019.pdf

50. Boelee E, Geerling G, van der Zaan B, Blauw A & Vethaak AD. Water and health: From environmental pressures to integrated responses. Acta Tropica. 1;193:217–26. 2019. https://doi.org/10.1016/j.actatropica.2019.03.011

51. Jambeck JR, Geyer R, Wilcox C, Siegler TR, Perryman M, Andrady A, Narayan M & Law KL. Plastic waste inputs from land into the ocean. Science. 347(6223):768-71. 2015. https://www.science.org/doi/abs/10.1126/science.1260352#:~:text=DOI%3A%2010.1126/science.1260352

52. Wabnitz C & Nichols WJ. Editorial: Plastic Pollution: An Ocean Emergency. Marine Turtle Newsletter.129:1-3. 2010. https://www.researchgate.net/profile/Wallace-Nichols/publication/268187066_Editorial_Plastic_Pollution_An_Ocean_Emergency/links/54c622550cf256ed5a9c8f3c/Editorial-Plastic-Pollution-An-Ocean-Emergency.pdf

53. Schwabl P, Köppel S, Königshofer P, Bucsics T, Trauner M, Reiberger T, et al. Detection of Various Microplastics in Human Stool. Annals of Internal Medicine. 171(7):453.

2019. https://doi.org/10.7326/M19-0618

54. Oliveri Conti G, Ferrante M, Banni M, Favara C, Nicolosi I, Cristaldi A, et al. Micro- and nano-plastics in edible fruit and vegetables. The first diet risks assessment for the general population. Environmental Research. 187(187):109677. 2020. https://doi.org/10.1016/j.envres.2020.109677

55. Diaz-Basantes MF & Conesa JA, Fullana A. Microplastics in honey, beer, milk and refreshments in Ecuador as emerging contaminants. Sustainability. 12(14):5514. 2020.https://doi.org/10.3390/su12145514

56. Peixoto D, Pinheiro C, Amorim J, Oliva-Teles L, Guilhermino L & Vieira MN. Microplastic pollution in commercial salt for human consumption: A review. Estuarine, Coastal and Shelf Science. 219:161–8. 2019. https://doi.org/10.1016/j.ecss.2019.02.018

57. Lee HJ, Song NS, Kim JS & Kim SK. Variation and uncertainty of microplastics in commercial table salts: Critical review and validation. Journal of Hazardous Materials. 402:123743. 2021. https://doi.org/10.1016/j.jhazmat.2020.123743

58. Kedzierski M, Lechat B, Sire O, Le Maguer G, Le Tilly V & Bruzaud S. Microplastic contamination of packaged meat: Occurrence and associated risks. Food Packaging and Shelf Life. 24:100489. 2020. https://doi.org/10.1016/j.fpsl.2020.100489

59. Winkler A, Santo N, Ortenzi MA, Bolzoni E, Bacchetta R & Tremolada P. Does mechanical stress cause microplastic release from plastic water bottles? Water Research. 166(0043-1354):115082. 2019. https://doi.org/10.1016/j.watres.2019.115082

60. Ramage SJFF, Coull M, Cooper P, Campbell CD, Prabhu R, Yates K, et al. Microplastics in agricultural soils following sewage sludge applications: Evidence from a 25-year study. Chemosphere. 376:144277. 2025. https://doi.org/10.1016/j.chemosphere.2025.144277

61. Cox KD, Covernton GA, Davies HL, Dower JF, Juanes F & Dudas SE. Human consumption of microplastics. Environmental Science & Technology. 53(12):7068–74. 2019. https://pubs.acs.org/doi/abs/10.1021/acs.est.9b01517

62. Mercogliano R, Avio CG, Regoli F, Anastasio A, Colavita G & Santonicola S. Occurrence of microplastics in commercial seafood under the perspective of the human food chain. A review. Journal of Agricultural and Food Chemistry. 68(19):5296–301. 2020. https://doi.org/10.1021/acs.jafc.0c01209

63. Wright SL & Kelly FJ. Plastic and human health: A micro issue? Environmental Science & Technology. 51(12):6634–47. 2017. https://doi.org/10.1021/acs.est.7b00423

64. Witczak A, Przedpełska L, Kamila Pokorska-Niewiada & Cybulski J. Microplastics as a threat to aquatic ecosystems and human health. Toxics. 12(8):571 -1. 2024. https://do i.org/10.3390/toxics12080571

65. Ward JE & Shumway SE. Separating the grain from the chaff: particle selection in suspension- and deposit-feeding bivalves. Journal of Experimental Marine Biology and Ecology. 300(1–2):83–130. 2004. https://doi.org/10. 1016/j.jembe.2004.03.002

66. Järlskog I, Strömvall AM, Magnusson K, Galfi H, Björklund K, Polukarova M, et al. Traffic-related microplastic particles, metals, and organic pollutants in an urban area under reconstruction. Science of The Total Environment.774:145503. 2021. https://doi.org/10.1016/ j.scitotenv.2021.145503

67. Dris R, Gasperi J, Rocher V, Saad M, Renault N & Tassin B. Microplastic contamination in an urban area: A case study in Greater Paris. Environmental Chemistry. 12(5):592. 2015. https://doi.org/10.1071/EN14167

68. Kacprzak S & Tijing LD. Microplastics in indoor environment: Sources, mitigation and fate. Journal of Environmental Chemical Engineering. 10(2):107359. 2022. https://doi.org/10.1016/j.jece.2022.107359

69. Ullah S, Ahmad S, Guo X, Ullah S, Ullah S, Nabi G, et al. A review of the endocrine disrupting effects of micro and nano plastic and their associated chemicals in

mammals. Frontiers in Endocrinology. 16;13. 2023. https://doi.org/10.3389/fendo.2022.1084236

70. Dalamaga M, Dimitrios Kounatidis, Dimitrios Tsilingiris, Vallianou NG, Karampela I, Sotiria Psallida, et al. The role of endocrine disruptors bisphenols and phthalates in obesity: Current evidence, perspectives and controversies. International journal of molecular sciences. 25(1):675–5. 2024. https://doi.org/10.3390/ij ms25010675

71. Ong HT, Samsudin H & Soto-Valdez H. Migration of endocrine-disrupting chemicals into food from plastic packaging materials: an overview of chemical risk assessment, techniques to monitor migration, and international regulations. Critical Reviews in Food Science and Nutrition. 1–23. 2020. https://doi.org/1 0.1080/10408398.2020.1830747

72. Zurub RE, Yusmaris Cariaco, Wade MG & Bainbridge SA. Microplastics exposure: implications for human fertility, pregnancy and child health. Frontiers in Endocrinology. 14. 2024. https://doi.org/10.3389/fendo.2023.1330396

73. Sharma RK, Kumari U & Kumar S. Impact of microplastics on pregnancy and fetal development: A systematic review. Cureus. 16(5). 2024. https://assets.cureus.c om/uploads/review_article/pdf/252793/20240521-4120 -1fyeuj1.pdf

74. Vincoff S, Schleupner B, Santos J, Morrison M, Zhang N,

Dunphy-Daly MM, et al. The known and unknown: Investigating the carcinogenic potential of plastic additives. Environmental Science & Technology. 58(24):10445–57. 2024. https://doi.org/10.1021/acs.est.3c06840

75. Irfan H, Irfan H, Khan MA, Inanc O & Hasibuzzaman MdA. Microplastics and nanoplastics: Emerging threats to cardiovascular health – a comprehensive review. Annals of Medicine and Surgery. 87(1):209–16. 2024. https://journals.lww.com/annals-of-medicine-and-surgery/fulltext/2025/01000/microplastics_and_nanoplastics___emerging_threats.33.aspx

76. Hu J, Qin X, Zhang J, Zhu Y, Zeng W, Lin Y, et al. Polystyrene microplastics disturb maternal-fetal immune balance and cause reproductive toxicity in pregnant mice. Reproductive Toxicology. 106:42–50. 2021. https://doi.org/10.1016/j.reprotox.2021.10.002

77. Senathirajah K, Attwood S, Bhagwat G, Carbery M, Wilson S & Palanisami T. Estimation of the mass of microplastics ingested – A pivotal first step towards human health risk assessment. Journal of Hazardous Materials. 404:124004. 2020. https://doi.org/10.1016/j.jhazmat.2020.124004

78. Amereh F, Babaei M, Eslami A, Fazelipour S & Rafiee M. The emerging risk of exposure to nano(micro)plastics on endocrine disturbance and reproductive toxicity: From a hypothetical scenario to a global public health

challenge. Environmental Pollution. 261:114158. 2020. https://doi.org/10.1016/j.envpol.2020.114158

79. Prata JC, da Costa JP, Lopes I, Duarte AC & Rocha-Santos T. Environmental exposure to microplastics: An overview on possible human health effects. Science of The Total Environment. 702(134455). 2020. https://doi.org/10.101 6/j.scitotenv.2019.134455

80. Rahman A, Sarkar A, Yadav OP, Achari G & Slobodnik J. Potential human health risks due to environmental exposure to nano- and microplastics and knowledge gaps: A scoping review. Science of The Total Environment. 757(143872):143872. 2021. https://doi.org/10.1016/j.scit otenv.2020.143872

81. Yee MSL, Hii LW, Looi CK, Lim WM, Wong SF, Kok YY, et al. Impact of microplastics and nanoplastics on human health. Nanomaterials. 11(2):496. 2021. https://doi.org/10.3390/nano11020496

82. Gou Z, Wu H, Li S, Liu Z & Zhang Y. Airborne micro- and nanoplastics: emerging causes of respiratory diseases. Particle and Fibre Toxicology. 21(1). 2024. https://doi.or g/10.1186/s12989-024-00613-6

83. Tang S, Wang M, Germ KE, Du HM, Sun WJ, Gao WM, et al. Health implications of engineered nanoparticles in infants and children. World Journal of Pediatrics. 11(3):197–206. 2015. https://doi.org/10.1007/s12519-015 -0028-0

84. Kumar R, Manna C, Padha S, Verma A, Sharma P, Dhar A, et al. Micro(nano)plastics pollution and human health: How plastics can induce carcinogenesis to humans? Chemosphere. 298(134267):134267. 2022. https://doi.org/10.1016/j.chemosphere.2022.134267

85. Dong CD, Chen CW, Chen YC, Chen HH, Lee JS & Lin CH. Polystyrene microplastic particles: In vitro pulmonary toxicity assessment. Journal of Hazardous Materials. 385:121575. 2020. https://doi.org/10.1016/j.jhazmat.2019.121575

86. Xiong X, Gao L, Chen C, Zhu K, Luo P & Li L. The microplastics exposure induce the kidney injury in mice revealed by RNA-seq. Ecotoxicology and environmental safety. 256:114821-1. 2023. https://doi.org/10.1016/j.ecoenv.2023.114821

87. Zhang Z, Chen W, Chan H, Peng J, Zhu P, Li J, et al. Polystyrene microplastics induce size-dependent multi-organ damage in mice: Insights into gut microbiota and fecal metabolites. Journal of Hazardous Materials. 461:132503-3. 2024. https://doi.org/10.1016/j.jhazmat.2023.132503

88. Prata JC. Airborne microplastics: Consequences to human health? Environmental Pollution. 234:115-26. 2018. https://doi.org/10.1016/j.envpol.2017.11.043

89. Campanale C, Massarelli C, Savino I, Locaputo V &

Uricchio VF. A detailed review study on potential effects of microplastics and additives of concern on human health. International Journal of Environmental Research and Public Health. 17(4):1212. 2020. https://dx.doi.org/1 0.13005/bpj/2959

90. Refay ASE, Armaneous AF, Salah DA, Youssef M, Salah E, Shady MA, et al. Phthalate exposure and pediatric asthma: A case control study among Egyptian children. Biomedical & Pharmacology Journal. 17(3):1489−98. 2024. https://dx.doi.org/10.13005/bpj/2959

91. Duh TH, Yang CJ, Lee CH & Ko YC. A study of the relationship between phthalate exposure and the occurrence of adult asthma in Taiwan. Molecules. 28(13):5230. 2023. https://doi.org/10.3390/molecul es28135230

92. Lavers, J. L., & Bond, A. L. Exceptional and rapid accumulation of anthropogenic debris on one of the world's most remote and pristine islands. Proceedings of the National Academy of Sciences, 114(23), 6052−6055. 2017. https://doi.org/10.1073/pnas.1619818114

93. Schuyler, Q., Hardesty, B. D., Wilcox, C., & Townsend, K. Global analysis of anthropogenic debris ingestion by sea turtles. Conservation Biology, 28(1), 129−139. 2013. https://doi.org/10.1111/cobi.12126

94. Rochman, C. M., Hoh, E., Kurobe, T., & Teh, S. J. Ingested plastic transfers hazardous chemicals to fish and induces

hepatic stress. Scientific Reports, 3(1). 2013. https://doi.org/10.1038/srep03263

95. Wilcox C, Van Sebille E & Hardesty BD. Threat of plastic pollution to seabirds is global, pervasive, and increasing. Proc Natl Acad Sci USA. 112(38):11899−904. 2015. https://doi.org/10.1073/pnas.1502108112

96. O'Hanlon NJ, James NA, Masden EA & Bond AL. Seabirds and marine plastic debris in the northeastern Atlantic: A synthesis and recommendations for monitoring and research. Environ Pollut. 231(Pt 2):1291−301. 2017. https://doi.org/10.1016/j.envpol.2017.08.101

97. Rapp DC, Youngren SM, Hartzell P & Hyrenbach KD. Community-wide patterns of plastic ingestion in seabirds breeding at French Frigate Shoals, Northwestern Hawaiian Islands. Mar Pollut Bull. 123(1−2):269−78. 2017. https://doi.org/10.1016/j.marpolbul.2017.08.047

98. Tanaka K, Takada H, Yamashita R, Mizukawa K, Fukuwaka MA & Watanuki Y. Accumulation of plastic-derived chemicals in tissues of seabirds ingesting marine plastics. Mar Pollut Bull. 69(1−2):219−22. 2013. https://doi.org/10.1016/j.marpolbul.2012.12.010

99. Tanaka K, Takada H, Yamashita R, Mizukawa K, Fukuwaka MA & Watanuki Y. Facilitated leaching of additive-derived PBDEs from plastic by seabirds' stomach oil and accumulation in tissues. Environ Sci Technol. 49(19):11799−807. 2015. https://doi.org/10.102

1/acs.est.5b01376

100. Duncan EM, Broderick AC, Fuller WJ, Galloway TS, Godfrey MH, Hamann M, et al. Microplastic ingestion ubiquitous in marine turtles. Glob Chang Biol. 25(2):744–52. 2019. https://doi.org/10.1111/gcb.14519

101. Unger B, Rebolledo ELB, Deaville R, Gröne A, Ijsseldijk LL & Leopold MF. Large amounts of marine debris found in sperm whales stranded along the North Sea coast in early 2016. Mar Pollut Bull. 112(1–2):134–41. 2016. https://doi.org/10.1016/j.marpolbul.2016.08.027

102. Mazurais D, Ernande B, Quazuguel P, Severe A, Huelvan C, Madec L, et al. Evaluation of the impact of polyethylene microbeads ingestion in European sea bass (Dicentrarchus labrax) larvae. Mar Environ Res. 112(Pt A):78–85. 2015. https://doi.org/10.1016/j.marenvres.2015.09.009

103. Duis, K., & Coors, A. Microplastics in the aquatic and terrestrial environment: sources (with a specific focus on personal care products), fate and effects. Environmental Sciences Europe, 28(1). 2016. https://doi.org/10.1186/s12302-015-0069-y

104. Amelia, T. S., Khalik, W. M., Ong, M. C., Shao, Y. T., Pan, H., & Bhubalan, K. Marine microplastics as vectors of major ocean pollutants and its hazards to the marine ecosystem and humans. Progress in Earth and Planetary Science, 8(1), 1-26. 2021. https://doi.org/10.1186/s40645-020-

00405-4

105. Gall, S., & Thompson, R. The impact of debris on marine life. Marine Pollution Bulletin, 92(1-2), 170-179. 2015. https://doi.org/10.1016/j.marpolbul.2014.12.041

106. Yorio P, Marinao C & Suarez N. Kelp Gulls (Larus dominicanus) killed and injured by discarded monofilament lines at a marine recreational fishery in northern Patagonia. Mar Pollut Bull. 85(1):186−9. 2014. https://doi.org/10.1007/978-3-030-20389-4_6

107. Prokić MD, Radovanović TB, Gavrić JP & Faggio C. Ecotoxicological effects of microplastics: Examination of biomarkers, current state and future perspectives. Trends Analyt Chem. 111:37−46. 2019. https://doi.org/10.1016/j.trac.2018.12.001

108. Prinz N & Korez Š. Understanding how microplastics affect marine biota on the cellular level is important for assessing ecosystem function: A review. In: Jungblut S, Liebich V, Bode-Dalby M, editors. YOUMARES 9 − The oceans: Our research, our future. Cham: Springer. 101−20. 2020. https://doi.org/10.1007/978-3-030-2038 9-4_6

109. Mejjad, N., Safhi, A. E. M., & Laissaoui, A. Insightful analytical review of potential impacts of microplastic pollution on coastal and marine ecosystem services. Journal of Hazardous Materials Advances, 17, 100578. 2025. https://doi.org/10.1016/j.hazadv.2024.100578

110. Romera-Castillo C, Pinto M, Langer TM, Alvarez-Salgado XA & Herndl GJ. Dissolved organic carbon leaching from plastics stimulates microbial activity in the ocean. Nat Commun. 9:1430. 2018. https://doi.org/10.1038/s41467-018-03798-5

111. Tetu SG, Sarker I, Schrameyer V, Pickford R, Elbourne LDH, Moore LR, et al. Plastic leachates impair growth and oxygen production in Prochlorococcus, the ocean's most abundant photosynthetic bacteria. Commun Biol. 2:184. 2019. https://doi.org/10.1038/s42003-019-0410-x

112. Cole, M., Lindeque, P., Halsband, C., & Galloway, T. S. Microplastics as contaminants in the marine environment: A review. Marine Pollution Bulletin, 62(12), 2588-2597. 2011. https://doi.org/10.1016/j.marpolbul.2011.09.025

113. Shen M, Ye S, Zeng G, Zhang Y, Xing L, Tang W, et al. Can microplastics pose a threat to ocean carbon sequestration? Mar Pollut Bull. 150:110712. 2020. https://doi.org/10.1016/j.marpolbul.2019.110712

114. Wang F, Wang F & Zeng EY. Chapter 7-Sorption of toxic chemicals on microplastics. In: Zeng EY, editor. Microplastic contamination in aquatic environments. Elsevier; 225–47. 2018. https://doi.org/10.1016/B978-0-443-15332-7.00011-9

115. Galgani, L., & Loiselle, S. A. Plastic pollution impacts on marine carbon biogeochemistry. Environmental Pollution, 268, 115598. 2020. https://doi.org/10.1016/j.envpol.2020.115598

116. Sharma S, Sharma V & Chatterjee S. Microplastics in the Mediterranean Sea: Sources, pollution intensity, sea health, and regulatory policies. Front Mar Sci. 8:634934. 2021. https://doi.org/10.3389/fmars.2021.634934

117. Schuyler, Q. A., Wilcox, C., Townsend, K. A., Wedemeyer-Strombel, K. R., Balazs, G., Van Sebille, E., & Hardesty, B. D.. Risk analysis reveals global hotspots for marine debris ingestion by sea turtles. Global Change Biology, 22(2), 567–576. 2015. https://doi.org/10.1111/gcb.13078

118. McIlgorm A, Raubenheimer K & McIlgorm DE. Update of 2009 APEC report on economic costs of marine debris to APEC economies. Wollongong: University of Wollongong. 2020. https://www.apec.org/docs/default-source/Publications/2020/3/Update-of-2009-APEC-Report-on-Economic-Costs-of-Marine-Debris-to-APEC-Economies/220_OFWG_Update-of-2009-APEC-Report-on-Economic-Costs-of-Marine-Debris-to-APEC-Economies.pdf

119. Organisation for Economic Co-operation and Development (OECD). Global Plastics Outlook Policy Scenarios to 2060. Paris: OECD. 2022. https://read.oecd-ilibrary.org/view/?ref=1143_1143481-88j1bxuktr&title=Global-

Plastics-Outlook-PolicyScenarios-to-2060-Policy-Highlights.

120. Riegl, B., & Piller, W. E. Possible refugia for reefs in times of environmental stress. International Journal of Earth Sciences, 92(4): 520–531. 2003. https://doi.org/10.1007/s00531-003-0328-9

121. Ferraro G & Failler P. Governing plastic pollution in the oceans: Institutional challenges and areas for action. Environmental Science & Policy. 112:453-60. 2020. https://doi.org/10.1016/j.envsci.2020.06.015

122. Tessnow-von Wysocki I & Le Billon P. Plastics at sea: Treaty design for a global solution to marine plastic pollution. Environmental Science & Policy. 100:94-104. 2019. https://doi.org/10.1016/j.envsci.2019.06.005

123. Da Costa JP, Mouneyrac C, Costa M, Duarte AC & Rocha-Santos T. The role of legislation, regulatory initiatives and guidelines on the control of plastic pollution. Frontiers in Environmental Science. 8:104. 2020. https://doi.org/10.3389/fenvs.2020.00104

124. Wu HH. A study on transnational regulatory governance for marine plastic debris: Trends, challenges, and prospect. Marine Policy. 136:103988. 2022. https://doi.org/10.1016/j.marpol.2020.103988

125. Secretariat of the Basel Convention. Basel convention on the control of transboundary movements of hazardous

wastes and their disposal. UNEP. 1989. https://we docs.unep.org/bitstream/handle/20.500.11822/8385/ -Basel%20Convention%20on%20the%20Control% 20of%20Transboundary%20Movements%20of%20H azardous%20Wastes%20-20113644.pdf?sequence=2& amp%3BisAllowed=

126. Basel Action Network. BAN is a core member of the global break free from plastics movement. BAN. https://www.b an.org/plastic-pollution-prevention

127. United Nations Environment Programme (UNEP). Global Plastics Treaty. UNEP 2022. https://www.unep.org/inc-plastic-pollution

128. Olatunji P, Potts J, Failler P & Austin R. Attitudes and perceptions of recreational boat owners on waste management processes: A case study of Chichester Harbour, United Kingdom. Journal of Sustainability Research. 2022 Sep 26;4(3):e220013. https://doi.org/10.20900/jsr20220013

129. International Maritime Organization (IMO). Garbage. London: International Maritime Organization. https://w ww.imo.org/en/OurWork/Environment/Pages/Garbage -default.aspx

130. International Maritime Organization (IMO). New action plan on marine litter, regarding the Annex V of the protocol of 1978 Relating to the International Convention for the Prevention of Pollution Form Ships, 1973. London,

UK. http://www.imo.org/en/MediaCentre/PressBriefin gs/Pages/20-marinelitteractionmecp73.aspx

131. Serra-Gonçalves C, Lavers JL, Tait HL, Fischer AM & Bond AL. Assessing the effectiveness of MARPOL Annex V at reducing marine debris on Australian beaches. Marine Pollution Bulletin. 191:114929. 2023. https://doi.org/ 10.1016/j.marpolbul.2023.114929

132. Wang S. International law-making process of combating plastic pollution: Status quo, debates and prospects. Marine Policy. 2023 Jan 1;147:105376. https://doi.org/ 10.1016/j.marpol.2022.105376

133. European Commission. Single-use plastics directive. Brussels: European Union. 2019. https://eur-lex.euro pa.eu/legal-content/EN/TXT/PDF/?uri=CELEX:32019L 0904#:~:text=The%20objectives%20of%20this%20Di rective,products%20and%20materials%2C%20thus% 20also

134. Cowan E, Booth AM, Misund A, Klun K, Rotter A & Tiller R. Single-use plastic bans: exploring stakeholder perspectives on best practices for reducing plastic pollution. Environments. 8(8):81. 2021. https://doi.org/ 10.3390/environments8080081

135. Kiessling T, Hinzmann M, Mederake L, Dittmann S, Brennecke D, Böhm-Beck M, Knickmeier K & Thiel M. What potential does the EU Single-Use Plastics Directive have for reducing plastic pollution at coastlines and

riversides? An evaluation based on citizen science data. Waste Management. 164:106-18. 2023. https://doi.org/10.1016/j.wasman.2023.03.042

136. Seas At Risk. Phasing out single-use plastics: how are EU countries performing? Brussels: Seas At Risk; 2022 Sept 21 https://seas-at-risk.org/press-releases/phasing-out-single-use-plastics-how-are-eu-countries-performing/

137. Kasznik D & Łapniewska Z. The end of plastic? The EU's directive on single-use plastics and its implementation in Poland. Environmental Science & Policy. 2023. 145:151-63.https://doi.org/10.1016/j.envsci.2023.04.005

138. Watkins E, Gionfra S, Schweitzer JP, Pantzar M, Janssens C & ten Brink P. EPR in the EU plastics strategy and the circular economy: A focus on plastic packaging. Institute for European Environmental Policy (IEEP). 2017. https://wm.turkmenistan.ecoline-int.org/wp-content/uploads/2021/08/EPR-and-plastics-report-IEEP-9-Nov-2017-final.pdf

139. Raubenheimer K, Urho N. Rethinking global governance of plastics–The role of industry. Marine Policy. 113:103802. 2020.https://doi.org/10.1016/j.marpol.2019.103802

140. Hart R. Shifting the burden of plastic bags: A proposal for a federal extended producer responsibility law. LSU J. Energy L. & Resources. 9:531. 2021. https://digitalco

mmons.law.lsu.edu/cgi/viewcontent.cgi?article=1224&context=jelr

141. Aiguobarueghian I, Adanma UM, Ogunbiyi EO & Solomon NO. Reviewing the effectiveness of plastic waste management in the USA. World Journal of Advanced Research and Reviews. 22(2):1720-33. 2024.https://doi.org/10.30574/wjarr.2024.22.2.1518

142. United Nations Environment Programme Banning Single-use plastics: Lessons and experiences from countries. UNEP. 2018. https://www.unep.org/resources/report/single-use-plastics-roadmap-sustainability

143. Kumar R, Verma A, Shome A, Sinha R, Sinha S, Jha PK, Kumar R, Kumar P, Shubham, Das S & Sharma P. Impacts of plastic pollution on ecosystem services, sustainable development goals, and need to focus on circular economy and policy interventions. Sustainability. 13(17):9963. 2021. https://doi.org/10.3390/su13179963

144. Bucknall DG. Plastics as a materials system in a circular economy. Philosophical Transactions of the Royal Society A. 2020 Jul 24;378(2176):20190268. https://doi.org/10.1098/rsta.2019.0268

145. Li H, Aguirre-Villegas HA, Allen RD, Bai X, Benson CH, Beckham GT, Bradshaw SL, Brown JL, Brown RC, Cecon VS & Curley JB. Expanding plastics recycling technologies: Chemical aspects, technology status and challenges. Green Chemistry. 24(23):8899-9002. 2022.

https://doi.org/10.1039/D2GC02588D

146. Lawal U & Valapa RB. Bioplastics: An introduction to the role of eco-friendly alternative plastics in sustainable packaging. Bio-based packaging: Material, Environmental and Economic Aspects. 24:319-34. 2021. https://doi.org/10.1002/9781119381228.ch18

147. Mangal M, Rao CV & Banerjee T.Bioplastic: An eco-friendly alternative to non-biodegradable plastic. Polymer International.72(11):984-96. 2023. https://doi. org/10.1002/pi.6555

148. Popp J, Lakner Z, Harangi-Rákos M & Fari M. The effect of bioenergy expansion: Food, energy, and environment. Renewable and Sustainable Energy Reviews. 32:559-78. 2014. https://doi.org/10.1016/j.rser.2014.01.056

149. Stevens ES. Green plastics: An introduction to the new science of biodegradable plastics. Princeton University Press; 2002. https://books.google.com.ng/books?hl=e n&lr=&id=AFO9Cajtv6EC&oi=fnd&pg=PR7&dq=Steven s+ES.+Green+plastics:+an+introduction+to+the+new +science+of+biodegradable+plastics.+Princeton+Univ ersity+Press%3B+2002.&ots=fcY0r_AI56&sig=dec6VY 2b30TCO-5_RtgPMftyT-Q&redir_esc=y#v=onepage& q=Stevens%20ES.%20Green%20plastics%3A%20an% 20introduction%20to%20the%20new%20science% 20of%20biodegradable%20plastics.%20Princeton% 20University%20Press%3B%202002.&f=false

150. Cho R. The truth about bioplastics. State of the planet. Columbia Climate School. 2017. https://news.climate.co lumbia.edu/2017/12/13/the-truth-about-bioplastics/

151. Lusher AL & Primpke S. Finding the balance between research and monitoring: When are methods good enough to understand plastic pollution? Environmental Science & Technology. 57(15):6033-9. 2023. https://doi. org/10.1021/acs.est.2c06018

152. Borg K, Lennox A, Kaufman S, Tull F, Prime R, Rogers L & Dunstan E. Curbing plastic consumption: A review of single-use plastic behaviour change interventions. Journal of Cleaner Production. 344:131077. 2022. https://doi.org/10.1016/j.jclepro.2022.131077

153. OECD. Policy scenarios for eliminating plastic pollution by 2040. OECD. 2023. https://www.oecd.org/en/pu blications/policy-scenarios-for-eliminating-plastic-pollution-by-2040_76400890-en.html

154. Prata, J.C., da Costa, J.P., Girão, A.V., Lopes, I., Duarte, A.C., & Rocha-Santos, T. Methods for sampling and detection of microplastics in water and sediment: A critical review. TrAC Trends in Analytical Chemistry, 110(150-159). 2019. https://doi.org/10.1016/j.trac.2018.1 0.029

155. Rochman CM. Microplastics research—from sink to source. Science. 360(6384):28−9. 2018. https://doi.org/ 10.1126/science.aar7734

156. Beaumont NJ, Aanesen M, Austen MC, Börger T, Clark JR, Cole M, et al. Global ecological, social and economic impacts of marine plastic. Mar Pollut Bull.;142:189–95. 2019. https://doi.org/10.1016/j.marpolbul.2019.03.022

157. Dasari VV & Suresh AK. The plastic problem in research laboratories: A call for sustainability. ACS Sustain Res Mgt. 2023. https://pubs.acs.org/action/showCitFormat s?doi=10.1021/acssusresmgt.5c00114

158. Borah SJ & Kumar V. Fundamental principles of waste management for a sustainable circular economy. In integrated waste management: A sustainable approach from waste to wealth. Singapore: Springer Nature Singapore; 1-11. 2024 https://doi.org/10.1007/978-981-97-0823-9_1

8

Acronyms

Below, please find a list of acronyms used in this book.

AFM Atomic Force Microscopy

AI Artificial Intelligence

AIIMS All India Institute of Medical Sciences

AMR Antimicrobial Resistance

AQ Air Quality

ARB Antibiotic-Resistant Bacteria

ASEAN Association of Southeast Asian Nations

ASPHER Association of Schools of Public Health in the European Region

BPA Bisphenol A

Bt Bacillus thuringiensis

CAPs Criteria Air Pollutants

CAQM Commission for Air Quality Management

Cas CRISPR-associated

CBD Convention on Biological Diversity

CCAC Climate and Clean Air Coalition

CCP Core Curriculum Programme

CETP Common Effluent treatment Plant

CGWB Central Ground Water Board

CO Carbon Monoxide

CO_2 Carbon Dioxide

COP Conference of Parties

CPCB Central Pollution Control Board

CRISPR Clustered Regularly Interspaced Short Palindromic Repeats

CSP Clean Seas Program

CVD Cardiovascular Disease

DALY Disability-Adjusted Life Years

DDT Dichlorodiphenyltrichloroethane

DNA Deoxyribonucleic Acid

DST Department of Science and Technology

Dte GHS Directorate General of Health Services

DWSC District Water Sanitation Committee

EAGs Expert Advisory Groups

ECDC European Centre for Disease Prevention and Control

EDCs Endocrine-Disrupting Chemicals

EF Emission Factors

EPA Environmental Protection Agency

EPC Empowered Programme Committee

EPR Extended Producer Responsibility

EU European Union

FAO Food and Agriculture Organization

FTIR Fourier Transform Infrared Spectroscopy

FTIR Fourier Transform Infrared

GCCHE Global Consortium on Climate Change and Health Education

GDP Gross Domestic Product

GHGs GreenHouse Gases

GLC Ground Level Concentration

GM Genetically Modified
GMC Genetically Modified Crop
GMO Genetically Modified Organism
GPA Global Program of Action
GPML Global Partnership on Marine Litter
GRAP Graded Response Action Plan
GURT Genetic Use Restriction Technology
HAS Health System Assessment
HCH Hexachlorocyclohexane
HIA Health Impact Assessment
HPAI Highly Pathogenic Avian Influenza
ICMR Indian Council of Medical Research
IEC Information, Education, and Communication
IFUD Instructions for Use and Disposal
IHME Institute of Health Metrics and Evaluation
IMD Indian Meteorological Department
IMO International Maritime Organization
IPCC Intergovernmental Panel on Climate Change
IWRM Integrated Water Resource Management
JJM *Jal Jeevan* Mission
MARPOL International Convention for the Prevention of
Pollution from Ships
MBBR Moving Bed Biofilm Reactor
MCES Marine and Coastal Ecosystem Services
MoEFCC Ministry of Environment, Forest & Climate Change
MoHFW Ministry of Health and Family Welfare
MP Microplastic
MtCO2e Megatons of Carbon Dioxide Equivalent
NAAQS National Ambient Air Quality Standards
NaCl Sodium Chloride
NAPCCHH National Action Plan on Climate Change and

Human Health

NCAP National Clean Air Programme

NCDC National Center for Disease Control

NCDs Non-Communicable Diseases

NDMA National Disaster Management Authority

NEERI National Environmental Engineering Research Institute

NGOs Non-Governmental Organizations

NIDM National Institute of Disaster Management

NIOH National Institute of Occupational Health

NMVOCs Non-Methane Volatile Organic Compounds

OCEMS Online Continuous Emission and Effluent Monitoring System

OH JPA One Health Joint Plan of Action

OHCC One Health Core Competencies

PCBs Polychlorinated Biphenyls

PGI Post Graduate Institute of Medical Education and Research

PHFI Public Health Foundation of India

PM Particulate Matter

POPs Persistent Organic Pollutants

REACH Registration, Evaluation, Authorization and Restriction of Chemicals

RSP Regional Seas Program

SBR Sequencing Batch Reactor

SCs Scheduled Castes

SDGs Sustainable Development Goals

SEM Scanning Electron Microscopy

SNOs State Nodal Officers

SOP Standard Operating Procedure

SPCB State Pollution Control Board

SPM Synthetic Polymer Microparticles
STs Scheduled Tribes
SUPD Single-Use Plastics Directive
TERI The Energy and Resources Institute
TESSy The European Surveillance System
UASB UpFlow Anaerobic Sludge Blanket
UNCLOS United Nations Convention on the Law of Sea
UNCLOS United nations Convention on the Law of the Sea
UNEA United Nations Environment Assembly
UNEP United Nations Environment Programme
USMCA United States–Mexico–Canada Agreement
UV Ultraviolet
VBDs Vector Borne Diseases
VWSC Village Water/Sanitation Committee
WHO World Health Organization
WOAH World Organisation for Animal Health
WQMIS Water Quality Management Information System
WQMS Water Quality Monitoring and Surveillance Framework
YBP Years Before Present
YP Young Professionals
YPP Young Professionals Programme

9

Meet the Authors

Editors

Dr. Saroj Pachauri, Public Health Specialist, Trustee, Center for Human Progress, New Delhi, India, and Director, POP (Protect Our Planet) Movement, New York, USA

Dr Pachauri, MD, PhD, DPH is currently focusing on research on health and climate change. As a public health physician, Dr. Pachauri has been extensively engaged with research on family planning, maternal and child health, sexual and reproductive health and rights, HIV and AIDS, and poverty, gender and youth. In 1996, she joined as Regional Director, South and East Asia, Population Council and established its regional office in New Delhi which she managed until 2014. In 2011, she was awarded the prestigious title of Distinguished Scholar, an honor rarely bestowed.

She worked with the Ford Foundation's New Delhi Office (1983-1994) and supported child survival, women's health, sexual and reproductive health, and HIV and AIDS programs. Before that, she worked with the International Fertility Research Program (IFRP) which was later renamed Family Health International (1971-1975) and the India Fertility Research Programme (1975-1983). She designed and monitored multi-centric clinical trials globally to assess the safety and effectiveness of fertility control technologies. During 1962-1971, as faculty of the Departments of Preventive and Social Medicine at the Lady Hardinge Medical College, New Delhi and the Institute of Medicine Sciences, Varanasi, she helped to develop this new discipline.

She secured the first position in the examination for the Diploma in Public Health. She was given the Lifetime Achievement Award by the Maharishri International University (MIU). She was declared Women of the Year in 2024 and 2025.

She has published twelve books of which four are on Climate Change and Health and contributed chapters to 20 books. She has over 100 publications in peer-reviewed journals and several articles in print media.

Dr. Ash Pachauri, Director, Center for Human Progress, New Delhi, India, and Senior Mentor, POP (Protect Our Planet) Movement, New York, USA

Dr. Ash Pachauri has a PhD in behavioral science and technology and a master's in international management. Having worked with McKinsey & Company before pursuing a career in the social development arena, Dr. Pachauri's experience in public health and sustainable development emerges from a range of initiatives. Notably, he has made significant contributions to the Bill & Melinda Gates Foundation by contributing to its public health and community agenda, the UN by focusing on youth, health, and the Sustainable Development Goals (SDGs), and the Center for Disease Control program interventions in the US by focusing on community interventions, especially for vulnerable youth. He has also been instrumental in founding and building the POP Movement and the World Sustainable Development Forum. He is a technical adviser to the World Health Organization on Self-Care Global Guidelines to support youth, communities, and global governments.

Dr. Pachauri has been a pioneer in the use of information technology for development. His innovative approaches have been key to spearheading community—and youth-led self-care interventions, leading to global capacity building and adoption of self-care among youth. As a master trainer in behavior change communications and strategic leadership, Dr. Pachauri has led over 20,000 workshops, events, and global outreach to youth and communities to promote global health and climate action.

Widely published, winner of the prestigious Overseas Research Scholarship, awarded for advanced studies in the U.K., and recognized for his academic achievements, Dr. Pachauri's awards and recognitions reflect his significant contributions to the field. The United Nations has recognized Dr. Pachauri for his dedication and leadership in their flagship publication, "Portraits of Commitment," A testament to his influence in the field. In 2021, he was awarded the GlobalMindED Inclusive Leadership Award for action in Energy and Sustainability, a recognition of his commitment to inclusive and sustainable development among young people worldwide. He is an Associate Fellow of the World Academy of Art and Science, a position that underscores his academic standing. Dr. Pachauri serves on the Boards and Advisory groups of several organizations and initiatives worldwide, including the global movement on bone health, the Climate Change Coalition, and the Global Union of Scientists for Peace. He demonstrates leadership and influence in the global health and climate action community.

Contributors

Dr. Karl F Conyard, ASPHER Core Curriculum Fellow, The Association of Schools of Public Health in the European Region (ASPHER), Brussles, Belgium; School of Public Health Physiotherapy and Sports Science, College of Health and Agriculture Sciences, University College Dublin, Ireland; Royal College of Surgeons in Ireland, University of Medicine and Health Sciences, Dublin, Ireland

Karl F. Conyard is a Core Curriculum Programme ASPHER Fellow based at the University College Dublin (UCD) School of Public Health Physiotherapy and Sports Science. Karl's work before joining ASPHER, was in the area of population health and epidemiology. He worked within Irelands' National Health Service (HSE) as an operations manager in the COVID-19 contact management program overseeing the national specialized clinical contact tracing team. Karl's original background was in auxiliary nursing. He then completed his BSc. in Health and Society at the Dublin City University (DCU) and attained his Master of Public Health (MPH) degree at

UCD. He is now on the mature entry medicine pathway at the Royal College of Surgeons in Ireland (RCSI) where he attained the RCSI Certificate in Lifestyle Medicine. Karl is a Fellow Member of the Royal Society for Public Health (UK) and the Royal Statistical Society (UK) and an Affiliate Member of the Royal College of Surgeons in Ireland. Karl's passion lies in epidemiology, infectious disease, competency and curriculum development, and practical One Health interventions.

Ms. Tara Chen, ASPHER Climate Health Fellow, The Association of Schools of Public Health in the European Region (ASPHER), Brussels, Belgium; Department of Geography and Environment Management, University of Waterloo, Canada

Tara Chen is a young public health professional working to strengthen health systems with a vision that "everything and everywhere is public health". She holds a BHSc from Western University, Canada as well as a dual Masters of Public Health (Europubhealth+) in Governance and Health Economics. She is currently a PhD candidate in the Department of Geography and Environmental Management, University of Waterloo, Canada, under the Geographies of Health in Place (GoHelP) lab. Tara served as a Climate-Health Fellow with the Association of Schools of Public Health in the European Region (ASPHER) from 2022-2024 where she worked on advocating for climate-health competencies in the public health curriculum. She has an active interest in health systems, nature-based social prescribing, one/planetary/eco-health, health literacy and cross-sectoral collaboration. Tara is also a strong advocate for amplifying young voices in dialogue, research, and practice.

Dr. Uma Divya Kudupudi, Core Curriculum Programme Assistant, School of Public Health Physiotherapy and Sports Science, College of Health and Agriculture Sciences, University College Dublin, Ireland

Dr. Uma Divya Kudupudi is an ASPHER Assistant at University College Dublin. Her expertise is in the field of epidemiology and infectious disease surveillance. Her background is in MBBS (Medicine and Surgery) and Master of Public Health from University College Dublin. Expanding to holding a position as an assistant to UCD ASPHER assisting with the core curriculum, Uma also works as Site Safety Representative for Microsoft enhancing occupational health. Uma is a Fellow of the Royal Society for Public Health (UK) and an Affiliate Member with the Royal College of Surgeons in Ireland. Uma's research interests include epidemiology, biostatistics, health promotion, administrative management, and publications.

Ms. Kirsten Duggan, ASPHER Young Professional, The Association of Schools of Public Health in the European Region (ASPHER), Brussels, Belgium

Kirsten Duggan is a researcher at the Gesundheitsamt Frankfurt am Main (public health authority), Germany. He earned a Master's Degree in Health Sciences from the University of Osnabrück (Germany) and a Master of Public Health at the University of Edinburgh, Scotland. The focus of her work is in climate change and health, particularly heat adaptation. With knowledge transfer being another focal point, she is not only part of the Climate and Health Working Group within ASPHER as a Young Professional, but also works with the

Core Curriculum for Public Health Programme Development group and the Digital Public Health Taskforce. Continuing her journey as a doctoral candidate, she is investigating climate inequality and vulnerability in the elderly population.

Dr. Parnian Jalili, PhD Candidate, School of Public Health Physiotherapy and Sports Science, College of Health and Agriculture Sciences, University College Dublin, Ireland

Dr. Parnian Jalili is a medical doctor who pursued a Master's in Public Health. She is currently a doctoral candidate in Public Health at University College Dublin. Her background is in biostatistics, epidemiology, One Health, and health advocacy. Her research focuses on wellbeing, career satisfaction, sustainable employability, and work engagement of Health Care Assistants (HCAs) and carers. She tutors in biostatistics, principles of epidemiology, data management, and advanced epidemiology modules for Master of Public Health students. Additionally, she tutors graduate and undergraduate medical students in introductory epidemiology, biostatistics, and public health.

Dr. Jwenish Kumawat, Research Fellow and PhD Candidate, School of Public Health Physiotherapy and Sports Science, College of Health and Agriculture Sciences, University College Dublin, Ireland

Dr. Jwenish Kumawat is a public health researcher and a PhD graduate from the UCD School of Public Health, Physiotherapy, and Sports Science. He has specialized in infectious disease epidemiology, health systems resilience, and post-pandemic preparedness. His work emphasizes interdisciplinary approaches within global health frameworks with a strong focus on One Health integration. He led pivotal studies on COVID-19 transmission dynamics, genome sequencing, and outbreak management directly informing public health responses in higher education settings. His contributions extend to curriculum development, embedding One Health principles into public health education to promote cross-sectoral collaboration, and future health system preparedness. Actively engaged with the ASPHER, Jwenish contributes to initiatives on pandemic preparedness, One Health, global health, criminal justice and public health, mental health, curriculum development, and public health capacity building across Europe. He holds a Master of Public Health (MPH) from UCD and a Bachelor of Dental Surgery (BDS) from the Sri Dharmasthala Manjunatheshwara (SDM) College of Dental Sciences, India. His work has been featured at leading global forums including the World Congress on Public Health and the European Public Health Conference.

Ms. Addiena Luke-Currier, PhD Candidate, Department of Sociology, School of Social Sciences and Philosophy, Trinity College Dublin, Ireland

Addiena Luke-Currier is a graduate of the Erasmus Mundus Joint Master's Program Europubhealth+ through which she received an MPH from University College Dublin and an MSc in European Public Health Leadership and Governance from the Maastricht University. She is currently a PhD student at Trinity College Dublin within the multidisciplinary One Health project RESIST-AMR. This project includes four PhD projects within the fields of clinical and environmental microbiology, botany, bioengineering, and sociology. She is studying the issue of antimicrobial resistance (AMR) in agricultural settings. Her project provides a social science perspective and aims to analyze agricultural stakeholders' practices and policies to identify institutional reform implications.

Ms. Marie Nabbe, ASPHER Young Professional, The Association of Schools of Public Health in the European Region (ASPHER), Brussels, Belgium; European Hospital and Healthcare Federation (HOPE), Brussels, Belgium

Marie Nabbe has been the EU Affairs Officer at the European Hospital and Healthcare Federation (HOPE) since 2021. She monitors EU national and regional policies and legislation on different topics through their impact on hospitals and healthcare such as cross-border threats and climate and environmental policies, both on mitigation and adaptation. Marie holds a Master's Degree in European studies from the Université Sorbonne Nouvelle (Paris, France) and a Master's Degree in Governance and Leadership in European Public Health from the Maastricht University (Maastricht, the Netherlands). She joined the ASPHER Young Professionals Programme in 2022 and participates in the Climate and Health Working Group. She is particularly interested in the links between climate, environment, and health. She strongly believes in collaboration between sectors to strengthen public health action.

Dr. Emrecan Ozeler, Student, Europubhealth+ European Public Health Master Program, University College Dublin,

Ireland

Dr. Emrecan Özeler is a second year student at the Europ-ubhealth+ European Public Health Master Program. After completing the first year of the program at University College Dublin (UCD), he specialized in biostatistics and epidemiology at Ecole des hautes études en santé publique (EHESP) in Paris. He is currently doing an internship on the global burden of foodborne diseases at the Technical University of Delft (TU Delft). Emrecan completed his undergraduate degree at the Ankara University Faculty of Veterinary Medicine and worked at the Ministry of Agriculture and Forestry of the Republic of Türkiye for more than 10 years as a European Union Specialist where his duties included the alignment of national legislation with EU legislation in the fields of animal health, plant health, and food and food safety as well as liaison with the European Commission's Directorate General for Health and Food Safety (DG SANTE) on sanitary and phytosanitary (SPS) issues.

Dr. Gerald Barry, Assistant Professor, School of Veterinary Medicine, College of Health and Agriculture Sciences, University College Dublin, Ireland; UCD One Health Centre, College of Health and Agriculture Sciences, University College Dublin, Ireland

Gerald Barry is an Assistant Professor of Virology in University College Dublin (UCD) and is Deputy Director of the UCD One Health Centre. Gerald completed his Master's and PhD at the University of Edinburgh. He then spent time in the Roslin Institute in Edinburgh before moving to the Centre for Virus Research in the University of Glasgow. In 2015, Gerald joined

UCD and since then has led a research group that studies virus infections in both animals and humans. Gerald is particularly interested in the host immune system and how that interacts with a virus during an infection. While promoting and developing a deeper understanding of One Health. Gerald sits on the Council of the Microbiology Society and is President of the Irish Association for Veterinary Teaching and Research Work (AVTRW).

Dr. Laurent Chambaud, ASPHER Climate Health Lead, The Association of Schools of Public Health in the European Region (ASPHER), Brussles, Belgium

Prof. Dr. Laurent Chambaud is the ASPHER Lead on Climate and Health and former Dean of the EHESP School of Public Health in Rennes (France). As a public health physician, he held key positions in public health policy, education, and international collaboration. Beginning in maternal and child health, he spent nearly eight years in the public health network in the Province of Quebec (Canada). He later trained as a medical inspector and became a national expert at the European Commission launching the first European communicable diseases surveillance decision. Member of the Social Affairs General Inspectorate, he also served as Director of Public Health at Île-de-France's Regional Health Agency and became advisor to the Minister of Health. Laurent has led training, research, and global collaboration in ASPHER and APHEA (Agency for Public Health Education Accreditation). Most recently, he facilitated scholarship exchanges for Ukrainian public health students. Awarded the 2022 Andrija Štampar Medal, he continues to shape international public health leadership.

Dr. Mary B Codd, ASPHER Core Curriculum Programme Lead, Associate Professor of Epidemiology & Biostatistics, School of Public Health Physiotherapy and Sports Science, College of Health and Agriculture Sciences, University College Dublin, Ireland; The Association of Schools of Public Health in the European Region (ASPHER), Executive Board, Brussles, Belgium

Prof. Mary Codd is Associate Professor of Epidemiology and Biostatistics at the UCD School of Public Health Physiotherapy and Sports Science. She is Director of the Master of Public Health (MPH) Program and the Erasmus Mundus Joint European Master of Public Health (Europubhealth) at UCD. She graduated in medicine from University College Dublin. Following a residency in internal medicine at the Mater Misericordiae Hospital, Dublin and Middlesex Hospital, London. She trained in epidemiology and biostatistics at the Mayo Clinic, Rochester, Minnesota, and the School of Public Health, University of Minnesota. Mary is a fellow of the Royal College of Physicians in Ireland, Faculty of Public Health Medicine. As a member of the Executive Board of ASPHER (Association of Schools of Public Health of the European Region), Mary was Project Scientific Lead on the ECDC Core Competencies in Applied Infectious Disease Epidemiology (2020-22) and is now the Lead of the ASPHER Core Curriculum Programme.

Mr. Patrick Wall, Professor of Public Health, School of Public Health Physiotherapy and Sports Science, College of Health

and Agriculture Sciences, University College Dublin, Ireland

Patrick Wall is a Professor of Public Health at University College Dublin (UCD) and a member of the University College Dublin, National University of Ireland (NUID UCD) Institute for Food and Health. He holds degrees in Veterinary and Human Medicine, an MSc in Infectious Diseases, an MBA, and a Diploma in Corporate Governance. A specialist in food safety, nutrition, and disease prevention, Patrick was the first Chief Executive of the Food Safety Authority of Ireland (FSAI) and later chaired the European Food Safety Authority (EFSA). He advised on food safety for the 2008 Beijing Olympics, the Chinese National Centre for Food Safety Risk Assessment, and the Saudi Food and Drug Authority. His research focuses on food risk communication, One Health, and behavioral change in disease prevention. He currently leads EU-China collaborations on food safety and chairs Ireland's Campylobacter Stakeholders Group working to reduce foodborne illness.

Ms. Komal Mittal, Research Associate, Center for Human Progress, New Delhi, India and Youth Mentor, POP (Protect Our Planet) Movement, New York, USA

Komal Mittal is a researcher, mentor, and advocate committed to driving social change through research, education, and youth empowerment. She is a Research Associate at the Center for Human Progress and a Global Youth Mentor with the Protect Our Planet (POP) Movement. Currently, she is pursuing her PhD in Sociology. Her research focuses on understanding self-care practices among the most marginalized and vulnerable communities in India.

With a deep passion for public health and sustainable development, Komal led a national youth group under the Joint United Nations Programme on HIV/AIDS promoting leadership and advocacy for the Sustainable Development Goals. She actively participated in and organized numerous national and international conferences, presenting research on public health, climate change, and human rights. She was recognized with the "Research Excellence Award" in Biotechnology for her work on the extraction of acid-soluble collagen from soybean and tomato in 2020.

Komal co-authored the book Sexual and Reproductive Health and Rights: Self-care for Achieving Universal Health Coverage with Dr. Saroj Pachauri and Dr. Ash Pachauri, published by Springer in 2020. Her research contributions extend to climate change, mental health, and the struggles of vulnerable populations, particularly in the wake of the COVID-19 pandemic. Through her work, she shed light on how marginalized communities navigate systemic deprivation and limited access to basic health and human rights.

Komal is deeply committed to environmental conservation and planetary wellbeing. She believes in the power of education, awareness, and community-driven action to foster a more sustainable future. By empowering young leaders, she champions a message of resilience, advocacy, and transformation—leading with unwavering dedication, perseverance, and hope for a better tomorrow.

Ms. Drishya Pathak, Project Management Consultant for Central Drug Standard Control Organization with the Bill & Melinda Gates Foundation

Drishya Pathak is a public health professional with eight years of experience in the public health and development sectors. She completed her Masters Degree in Health Management from the International Institute of Health Management and Research in 2019 and her undergraduate degree in Microbiology from the University of Delhi. She received the Alexander Von Humboldt Climate Protection Fellowship in 2023. She has presented several reports in the area of public health. She supported the planning, implementation, and documentation of the Second World Sustainable Development Forum in Durango, México in 2020.

From 2019-2024, Drishya worked as a Research Associate with the Center for Human Progress. She worked on projects on the sexual and reproductive health needs of key populations and people living with HIV (PLHIV). She has hands-on experience of working closely on issues like acudetox, education and awareness, and gender empowerment. She was involved with the implementation and training for the Integrated Digital Adherence Technology (IDAT) Project on Tuberculosis. She approaches environmental sciences through a public health lens as demonstrated in her recent work on 'Leaching of Chemicals from Plastic Food Contact Materials into Food', which was recognized on an international platform.

Mr. Ivan Ransom Rodriguez, Sustainability Ambassador and Mentor, POP (Protect Our Planet) Movement, New York, USA

Ivan Ransom Rodríguez is a sustainability specialist with a robust academic background and over ten years of experience in climate advocacy and environmental solutions. He holds a BSc in Biology from the Universidad Nacional Autónoma de México (UNAM) and has specialized in Sustainable Development. He holds a postgraduate diploma from the University of St Andrews. Ivan's commitment to global wellbeing is evident in his diverse contributions, from co-authoring two research papers on cellular and molecular biology to conducting cancer research at the University of Aberdeen.

As a climate advocate, Ivan has been instrumental in raising awareness about environmental issues and implementing practical solutions. His innovative approach, grounded in scientific research, has enabled him to successfully win public and private sector competitions. Ivan is also the Co-Founder of Resilience 2020, a company based in a circular economy that generates income for households and farmers while reducing their carbon footprint.

Ivan has represented the Protect Our Planet Movement at numerous international events contributing significantly to the organization through both virtual and in-person events across different countries. In addition to his professional achievements, he is a Peace Ambassador for the Institute for Economics and Peace. Ivan's comprehensive understanding of sustainability and resilience continues to drive impactful change in the environmental sector.

Mr Tariq Ahmad, Junior Research Associate, Resilience and Sustainability of Ocean Resources Cluster, National

Maritime Foundation, India

Mr Tariq Ahmad is a Junior Research Associate at the National Maritime Foundation (NMF) working on climate change adaptation and resilience. He is an architect-planner by profession and has academically worked on riverine port design and urban and district development plans as well as with hazard risk vulnerability assessment plans. He has worked professionally in green building ratings, building physics analyzes, and research and development of sustainable cooling systems. At the National Maritime Foundation (NMF), he is primarily working on port climate change adaptation, maritime spatial planning (MSP), and nature-based solutions (NbS).

Dr. Chime Youdon, Research Fellow, Head, Resilience and Sustainability of Ocean Resources (RSOR) Cluster, National Maritime Foundation, India

Dr Chime Youdon is a Research Fellow and the Head of the Resilience and Sustainability of Ocean Resources (RSOR) Cluster at the National Maritime Foundation in New Delhi. She focuses on research aimed at improving resilience for coastal communities and maritime infrastructure in the face of climate challenges. Currently, she is leading a collaborative project with Keio University and the Coalition for Disaster Resilient Infrastructure (CDRI) to develop a risk assessment model for seaport ecosystems with the goal of strengthening critical infrastructure. Dr Youdon advocates for sustainability through nature-based solutions and supports the development of the blue economy while ensuring maritime security in the changing climate landscape of the Indo-Pacific. She actively contributes to various expert working groups including the QUAD Working Group on Seaport Resilience with CDRI. She has received several prestigious fellowships, such as the CDRI Fellowship (2021–2022), the Admiral KK Nayyar Fellowship (2020–2023), and the Secure and Sustainable Infrastructure Resilience Program, under QUAD Fellowship by the CEPT (Centre for Environmental Planning and Technology) Research & Development Foundation (CRDF) and U.S. State Department (2024). Dr Youdon is also an author of a recent book Rising Seas and Coastal Impacts: Metropolitan Resilience in India that offers insights into resilience strategies for megacities facing coastal risks.

Mr. Soham Agarwal, Associate Fellow, Resilience and Sustainability of Ocean Resources (RSOR) Cluster, National Maritime Foundation, India

Mr. Soham Agarwal is an Associate Fellow in the Resilience and Sustainability of Ocean Resources (RSOR) Cluster at the National Maritime Foundation (NMF).His areas of research include the seabed (particularly the legal aspects thereof) including in the context of seabed warfare. Critical maritime infrastructure such as submarine communication cables and deep-seabed ocean resources extraction has formed a central part of his research. He has published on these subjects and has presented at both national and international Track 1 and Track 1.5 forums which include the Indian Navy, the National Security Council Secretariat, the Maritime Law Workshop of the Colombo Security Conclave, and the International Cable Protection Committee's symposium on undersea cable resilience in Valentia Island, Ireland. He was invited as a Defence Special Invitee by the Australian Department of Defence where he presented options for Indo-Australia collaboration on undersea infrastructure resilience. He is a recipient of the prestigious Coalition for Disaster Resilient Infrastructure (CDRI) fellowship 2024-2025 to enhance disaster-resilience of undersea communication cable infrastructure in India and an Infrastructure Resilient Island States' project on seaport resilience in Seychelles.

He received an LLB (Hons.) from the University of Nottingham, United Kingdom and completed an LLB Bridge Course in Indian law at the National Law University of Delhi. His experience includes research, drafting, and advocacy at law firms and counsel chambers in India. As a true enthusiast of the underwater domain, he also has an Open Water Diver certification.

Dr Gulshan Sharma, Associate Fellow, Resilience and Sustainability of Ocean Resources (RSOR) Cluster, National Maritime Foundation, India

Dr Gulshan Sharma is an environmental scientist specializing in climate change resilience, blue economy, and marine plastic pollution. At the National Maritime Foundation, Dr Sharma is working on blue economy and climate change related issues with the Resilience and Sustainability of Ocean Resources (RSOR) Cluster. She served as a Research Associate in the Blue Economy Centre for Excellence at the International Centre for Environment Audit and Sustainable Development (iCED), under the Comptroller and Auditor General of India. As a council member of the Indian Meteorological Society and the American Chemical Society, Dr Sharma engages with scientific communities. Dr Sharma is committed to advancing research on environmental science and sustainability.

Mr. Paul Ayomide Eweola, Founder and Executive Director, Aquaworld Community Development, Nigeria

Paul Ayomide Eweola is a marine scientist with a strong passion for ocean conservation and climate resilience. He is an MSC student in Marine Environment at the prestigious Erasmus Mundus Joint Master's Degree (EMJMD) funded by the European Union's Erasmus+ Scholarship. He holds a Bachelor of Technology in Fisheries and Aquaculture from the Federal University of Technology, Akure (FUTA), Nigeria where he conducted research on toxic proteomic, environmental impact assessment, and aquatic ecosystem restoration.

Eweola is the Founder and Executive Director of Aquaworld, a youth-led organization dedicated to enhancing coastal community resilience in Nigeria and promoting marine biodiversity protection. Through this platform, he leads research-driven initiatives such as the Ocean Leaders Drive Project (funded by the National Geographic Society) and the Fishers Prep Initiative (supported by the Sustainable Ocean Alliance and Environmental Defense Fund) which focus on marine plastic pollution, climate adaptation, and sustainable fisheries.

In addition to his leadership at Aquaworld, Eweola serves as the Director of Operations at the U-Recycle Initiative Africa where he oversees projects that promote circular economy and environmental sustainability across Africa. His impact extends globally—he is a Young National Geographic Explorer, an Earth Day Network Youth Ambassador (Plastic Champion), a two-term alumnus of the World Ocean Day Youth Advisory Council, and a member of the UN Ocean Decade's Western Tropical Atlantic Clean Ocean Working Group.

As a recognized thought leader in ocean and climate advocacy, Eweola has spoken at several international events including COP26, the Africa Blue Economy Week, the Youth Connekt Africa Summit (Kenya), and Our Global Ocean Conference (Greece). His contributions to SDG 14 (Life Below Water) have earned him multiple honors including recognition by the Nigeria Youth SDG Network and Oxfam Nigeria and inclusion in Nigeria's first-ever SDG Compendium (2023).

Beyond research and advocacy, Eweola has secured over $240,000 in grants, scholarships, and funding for various projects and academic pursuits. His work exemplifies the intersection of scientific research, policy engagement, and grassroots action to drive meaningful change in marine conservation and climate resilience.

Mr. Kehinde Adejugbagbe, Research Advisor, Aquaworld Community Development, Nigeria

Kehinde is an ecotoxicologist whose work is focused on safe fish production, a contaminant/pollutant-free environment, and a sustainable fish farming business. His research and writing are entrenched in the quantitative assessment of contaminants/pollutants (legacy and emerging), differential build-up and its correlational public health risks.

His recent research is focused on aquatic ecotoxicology through investigation of contaminant/pollutant fate, transmutation and interaction within environmental biota and non-biota components upto detailed levels of organization and remediation. He is a mentor and advisor to individual environmental/ocean organizations and a technical partner of agro-fishery business ventures.

Mr. Paul Olatunji, Foreign Hub Lead, Aquaworld Community Development, Nigeria

Paul Olatunji is an aquatic pollution expert and ecotoxicologist who has a strong research focus on understanding and mitigating anthropogenic impacts on aquatic ecosystems, particularly plastic pollution. His work is dedicated to leveraging scientific research to address critical environmental challenges in marine and freshwater systems. He has extensive research experience. He has four notable publications in peer-reviewed journals and presentations at international conferences.

Paul holds an MSc in Coastal and Marine Resource Management where he investigated boat owners' attitudes and perceptions toward waste management in Chichester Harbour. Additionally, as an Erasmus MSc graduate in Marine Environment and Resources, he explored the interactions of freshwater bivalves with microplastics of varying polymer types, sizes, and shapes.

Paul is deeply committed to environmental advocacy. As the Foreign Hub Lead of Aquaworld, a renowned youth-led organization focused on aquatic conservation, he actively engages in volunteer initiatives and awareness campaigns to combat aquatic pollution and promote a cleaner and healthier ocean.

Mr. Mustapha Suebat, Member, Aquaworld Community Development Initiative, Nigeria

Suebat Oluwakemi Mustapha is a dedicated environmental scientist driven by a passion for conserving and restoring aquatic ecosystems. Currently advancing her expertise as an Erasmus Mundus scholarship recipient in the Master's in Applied Ecohydrology (MAEH) program at the University of Antwerp Belgium, she investigates pressing issues such as aquatic habitat degradation. Her research focuses on ecosystem-based adaptation strategies for aquatic environments, leveraging tools like the R Programming language to develop geospatial ecosystem mapping techniques that identify pollution hotspots and prioritize targeted restoration measures.

Building on a foundational background in Fisheries and Aquaculture from the Federal University of Technology Akure (FUTA), Suebat has honed her interdisciplinary expertise in hydrology, watershed dynamics, and sustainable aquatic ecosystem management. She is deeply committed to bridging science, policy, and community engagement to address the intersection of environmental challenges and societal wellbeing. Beyond academia, Suebat actively contributes to global water governance and youth-led initiatives. She serves as a journal reviewer for the World Water Policy Journal evaluating research on equitable water resource management and is a member of the Ecohydrology Youth Network where she collaborates with emerging professionals to address hydromorphological challenges and advocate for nature-based solutions.

Ms. Simbiat Salaudeen, MSc Student, University of Basque Country, Spain

Simbiat Salaudeen is an early career researcher in the field of aquatic ecotoxicology. Her research focuses on the effects of pollutants on aquatic organisms and ecosystems. She has published a research article evaluating the effects of arsenic on the African catfish. Simbiat earned her Bachelor's Degree from the University of Ilorin and subsequently, she got a fully-funded Erasmus Mundus scholarship to embark on her Masters Degree. She is currently in her second semester at the University of the Basque Country, Spain. Simbiat is a One Health advocate who passionately advocates for responsible waste disposal
and chemical management.

Mr. Adetola Adebowale, Member, Aquaworld Community Development, Nigeria

Adetola Adebowale is an aquatic ecotoxicologist who studies the effects of pollution on aquatic ecosystems. His research focuses on assessing the ecotoxicological impacts of chemical and plastic contaminants in our environment. He has published two peer-reviewed articles assessing the bioavailability of contaminants in aquatic species. He earned his Master's Degree at the University of Cape Town where he investigated contaminants in sharks. He is currently furthering his studies in ecotoxicology at the Rheinland-Pfälzische Technische Universität (RPTU) Kaiserslautern-Landau in Germany. He is an active member of Aqua World, Early Career Ocean Professionals, and the Southern African Society of Aquatic Scientists where he contributes to a broader discourse and undertakes activities aimed at ensuring sustainable environmental protection, safeguarding biodiversity, and enhancing human health and safety.

Mr. Olagboye Olasunkanmi, Senior Program Manager, Aquaworld Community Development, Nigeria

Olagboye Olasunkanmi is a young geoscientist and researcher with a background in geochemistry and hydrogeology. He graduated with first-class honours in Applied Geology from the Federal University of Technology, Akure (FUTA), Nigeria. His research focuses on stable isotope geochemistry, hydrogeochemistry, and contaminant transport modelling, particularly in the context of sediment formation, fluid-rock interactions, and environmental sustainability. Olasunkanmi has contributed to peer-reviewed publications including studies on mineralogical composition, heavy metal distribution, and pollution indices in geologic formations.

As a recipient of the 2024 Geochemical Society Grant, he organized a stable Isotope Geochemistry Workshop bridging theoretical concepts with practical applications for emerging geoscientists. Olasunkanmi has held various leadership positions including serving as Senior Program Manager at Aquaworld CDI, where he led initiatives on climate adaptation, coastal resilience, and sustainable water resource management.

Ogunseitan Power, Research Associate, Aquaworld Community Development, Nigeria

Ogunseitan Power Peter holds an Honors Bachelor's Degree in Fishery Production from the Olabisi Onabanjo University, Ago Iwoye, Ogun State, Nigeria. With a strong foundation in environmental science, Peter's academic and professional journey has been deeply influenced by his commitment to sustainable practices and innovative solutions to environmental challenges.

Peter has cultivated a broad and insightful understanding of climate change and environmental safeguards through continuous learning and engagement with global experts. His passion for climate change and food transformation led him to undertake specialized training at the Wageningen Centre for Development Innovation in the Netherlands for expanding his knowledge and expertise in the field.

An ardent advocate for sustainability, Peter volunteers with Aquaworld Nigeria, where he serves as the Time-Out Team Coordinator focusing on aquatic service and sustainability. His role involves hands-on work on environmental management and community engagement to protect aquatic ecosystems.

His research on the physiochemical parameters and phyto-plankton distribution of Oyan Lake River, Ogun State con-tributed valuable data to the field of environmental science demonstrating his analytical and creative approach to address-ing complex environmental issues.

As a researcher and environmental advocate, Peter's core values of excellence, integrity, and professionalism guide his work. He believes in the transformative power of collective action for ocean health.